Developments in Atmospheric Science, 15

WEATHER AND CLIMATE OF THE ANTARCTIC

Developments in Atmospheric Science, 15

Weather and Climate of the Antarctic

W. SCHWERDTFEGER

Dr. phil. habil., Professor Emeritus of Meteorology

Department of Meteorology, University of Wisconsin, 1225 W. Dayton St., Madison, WI 53706, U.S.A.

ELSEVIER
Amsterdam — Oxford — New York — Tokyo   1984

ELSEVIER SCIENCE PUBLISHERS B.V.
Molenwerf 1,
P.O. Box 211, 1000 AE Amsterdam, The Netherlands

Distributors for the United States and Canada:

ELSEVIER SCIENCE PUBLISHING COMPANY INC.
52 Vanderbilt Avenue
New York, NY 10017

Library of Congress Cataloging in Publication Data

Schwerdtfeger, Werner.
    Weather and climate of the Antarctic.

    (Developments in atmospheric science ; 15)
    Bibliography: p.
    Includes index.
    1. Antarctic regions--Climate. I. Title. II. Series.
QC994.9.S33 1984   551.6998'9   83-27529
ISBN 0-444-42293-5 (U.S.)

ISBN 0-444-42293-5 (Vol. 15)
ISBN 0-444-41710-9 (Series)

Printed in The Netherlands

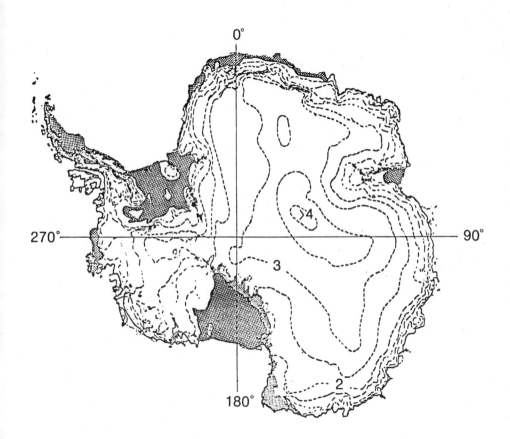

PREFACE

In the past two decades, considerable progress has been made in meteorological, climatological, and related glaciological research on the Antarctic and the surrounding ocean. A large part of the resulting literature refers directly or in a broader sense to climate problems. In contrast, little has been published about the vagaries of the weather in different sectors of the coasts of the continent and on its formidable high plateau, permanently snow and ice covered and yet one of the largest deserts on Earth. The thorough research reports and travelogs of the classical expeditions of Nordenskjöld, Scott, Shackleton, Mawson, and others some seventy years ago are still the best sources of information about what "bad weather" in the Antarctic really means.

The purpose of this book is to present an up-to-date description and, as far as possible, explanation of the meteorological characteristics of the southern continent, its inseparable topics weather and climate, including their close relation to the ice conditions on the surrounding waters. Some problems which need further field work and theoretical investigation are outlined in detail. Questions of weather forecasting, however, are discussed only where it appears that new information or understanding of the involved atmospheric processes make it worthwhile. The difficulties of an adequate interpretation of minor regional climate fluctuations --- physically explainable development versus chance --- are commented on. The text is thought to be understandable and usable not only for meteorologists and climatologists, but also for the much greater group of learned or learning natural scientists who are interested in the polar regions. Therefore, in the first sections of several chapters some general remarks are included that are not needed by professional meteorologists; lengthier mathematical considerations have been moved into two appendices.

The temperature regimes in different parts of the Antarctic can be important for investigations in various branches of the polar natural sciences, but complete data series of many stations are not easily available. Therefore, two tables in the Appendix give a data collection which, I think, is more useful than mean maps. I have also tried to enclose a fair survey of the sources, books, and articles consulted. Wherever appropriate, I have added short excerpts from original publications of years long past. By no means, however, could I claim to have treated all new and valuable papers adequately; I apologize to the colleagues whose fine work might not be duly referred to. Still, about 50% of all scientific journal articles quoted are less than ten years old. I hope this book will help to incite younger scientists to attack one or more of the many problems which the antarctic atmosphere offers. After 52 active years in various branches of meteorology and four continents, I can assure them it is a worthwhile undertaking.

ACKNOWLEDGEMENTS

I am indebted to my colleagues C. Bentley, D. Bromwich, M. Kuhn, D. Limbert, M. Savage, F. Sechrist, C. Stearns, and several commanding officers of the Argentine Navy. All these good friends helped either with some hard-to-get data and reports, or by describing to me the working conditions and other details of many Antarctic operations and places never seen by myself. My former students, Tom Parish and David Martin, critically reviewed the draft for chapters 3 and 4. The National Meteorological Services of Australia, New Zealand, and South Africa, and the British Antarctic Survey, openhandedly provided valuable information and publications. The text was revised by John Stremikis, most of the graphs drawn by Ms. M. Sievers, all the typing by Mrs. E. Singer. In the seventeen years before I started to prepare this book, the National Science Foundation, Division of Polar Programs, supported my work and the work of my students on problems of polar meteorology. All this generous help is gratefully acknowledged.

CONTENTS

VARIOUS ABBREVIATIONS, SYMBOLS, AND UNITS

1. The names of meteorological stations are written in capital letters throughout, in order to discern them from other geographical names. Abbreviations
   used are shown in Appendix A 1.

2. Where long series of observations or measurements are given or quoted, $\underline{i}$
   stands for "with interruptions".

3. When the time of observations is indicated, z means Greenwich mean time
   (GMT); L.T. = local time.

4. Wherever there could be any doubt whether air temperatures, ice temperatures,
   or surface water temperatures are meant, the notations T_a, T_i, or T_w are
   used. Only in Section 3.6 T_w refers to windchill.

5. The subscript small s stands for surface, subscript capital S for shortwave
   (radiation).

6. A dot on top of the equal sign = means "approximately equal".

7. Monthly, seasonal, or annual sums of radiative energy fluxes are given in

   $MJ\ m^{-2}$ (Megajoule per m^2).
   $1\ MJ\ m^{-2} = 10^6 J\ m^{-2} = 23.9\ \ell y = 23.9\ cal\ cm^{-2}$.

8. All temperatures are measured in °C, only potential temperatures in °K;
   all pressure values in mb, 1 mb = 100 Pascal; all winds in meter per second,
   1 m/sec = 2 knots; where more accuracy is justified, 1 m/sec = 1.94 knots =
   2.24 statute miles per hour = 3.60 km per hour.

Chapter 1.

INTRODUCTION

Out on the ice shelf the Sun had been shining in the morning. Then, a thick low overcast moved in with warmer and moister air from the north, and the visible world became invisible. Light was diffusively reflected from the cloud deck above as much as from the surface below, the sky blended into the snow cover. There was no horizon, no shadow, no perception of depth or height above surface, or of obstacles like sastrugi and snow waves. A great white nothing: Whiteout. An explorer's torment, a pilot's curse, a meteorologist's weather.

Another place, another time: Back home in Australia, he would never have dreamed of anything like what he had experienced, and some times barely survived, in the past seventeen months. Now, at the end of the second July he spends at CAPE DENISON, meteorologist C.T. Madigan checks the records again. In the stormiest hour the mean windspeed amounts to 43 m/sec the average for the entire month is 25 m/sec; all this with temperatures far below freezing, and often snow-filled air. The mean wind speed for the two years 1912 and 1913 at CAPE DENISON is about 20 m/sec, at neighboring PORT MARTIN in 1950 and 1951 18 m/sec. These winds distinguish the climate of a small sector of the coast.

Obviously weather and climate in the Antarctic have their surprising traits, daring the observer in ice and snow, and also challenging the homebound admirer of the polar atmosphere who wants to show not only what happens, but why it happens. The word weather itself is here understood to mean "the state of the atmosphere and its short-time variations, mainly with respect to its effects upon life and human activities. Weather is thought of in terms of temperature, humidity, precipitation, cloudiness, visibility, and wind" (Huschke, 1959). The words weather in English and Wetter in German have the same meaning. In contrast, the same word in Russian, betep, just means "wind" and nothing else, suggesting the overwhelming importance of this one weather element for the survival in northeastern Europe, in times past; and in the Antarctic, still today. The word climate may be understood as a "synthesis of the weather" of a specific area (Huschke, ℓ.c.), represented by the statistical collective of the main meteorological variables during a lengthy interval of time, usually several decades. This latter limitation cannot yet be strictly applied to the Antarctic, or there would be "climate-data" for no more than two stations south of 60°S, and both north of the polar circle. Studies of the origin of the word climate lead back to the old Greeks. Ptolemy, astronomer at Alexandria, and the geographers of his time (second century A.D.) defined different zones according to the duration of the longest day or the maximum angular altitude of the Sun, its <u>incline</u> (Hann and Knoch, 1932). Since the Renaissance the meaning of the word has slowly developed to include any characteristic weather elements and their combination instead of sunshine alone. As far as <u>climate change</u> in recent times is concerned, it should be proclaimed only when significant varia- tions or trends appear in homogeneous, multiannual series of meteorological variables. For large time intervals and epochs long past, of course, geologi- cal and paleobiological evidence has to be established.

1.1 TOPOGRAPHY OF ANTARCTICA, AND EXTENT OF THE SURROUNDING ICE BELT

The continent is unique in various aspects. Less than 3% of its 14×10^6 km^2 are free of snow or ice at least in part of the year. The average elevation of Antarctica's surface is a little more than 2300 m, compared to Asia's, the second in line, 800 m. The high plateau of East-Antarctica has about 3.5×10^6 km^2 surface above 3000 m, and 3×10^4 km^2 (centered near 81°S, 77°E) above 4000 m, all snow-covered. On the major part of the plateau the slope of the terrain is so small, less than 1/500, that it can be seen only by means of a theodolite (Fig. 1.1). The Great Plains, between the Mississippi River and the eastern limit of the Rocky Mountains between 42 and 33°N lati- tude, has the same average slope. For many practical purposes such an incline

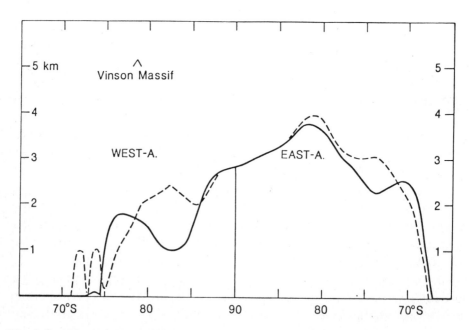

Fig. 1.1. Elevation profiles showing the contrast between West and East Antarctica. Solid line along 120°W and 60°E, dashed 100°W and 80°E; VINSON MASSIF at 85°W. Scale ratio: 1/558 (an angle of 45° on the graph is equivalent to an incline of 1/558).

is of little importance, but not so in meteorology. Under ordinary conditions in the lower troposphere, the slope of an isobaric surface is considerably smaller and yet quite significant.

In sharp contrast to the gentle grades of the east antarctic plateau, the points of greatest elevation have been found in the Ellsworth Mountains, rising out of the much lower west-antarctic plateau. The highest peak (5140 m), called Vinson Massif, lies at 78.6°S, 85.4°W, near the southern end of the Sentinel Range. These Alpine type mountains were first seen by Ellsworth and his pilot Hollick-Kenyon in 1935 from their single-engine airplane (Clinch, 1967).

With regard to problems of clean air and aerosol concentrations, it is of interest that there are at least five active volcanoes in the Antarctic. Best known is Mt. Erebus (3800 m), on Ross Island. Its smoke vane served the early expeditions of Shackleton and Scott to determine the prevailing direction of the wind near the 4000 m level. The second is Mt. Melbourne (2730 m) in the Transantarctic Mountains at 74.4°S, north of MCMURDO. The third volcanic spot active in recent years is on Deception Island (Baker et al., 1969; Orheim, 1970). Activity of two small volcanoes was observed in January 1982 some 220 km farther south in the group of the Seal Nunataks rising through the Larsen Ice Shelf near 65°S, 60°W (Gonzalez-Ferran, 1982). In the years 1962 to 1972 in which the station MATIENZO was in continuous operation, no volcanic activity has been reported from this region, but when Larsen discovered the Seal Nunataks in 1893 there was action on two other nunataks.

There are other features of the high southern latitudes to be kept in mind: The antarctic plateau is one of the two largest deserts of the world, the other one being, of course, the Sahara. The decision of which one deserves the doubtful honor to be number one, depends on definitions (Loewe, 1974). Referring to the extent of very small annual accumulation, M. Kuhn (1980) prefers the antarctic plateau which he knows so well. Griffith (1972), an expert on the climate of Africa, with arguments applicable to lower latitudes, favors the Sahara. As will be discussed in Chapter 5, almost the entire south polar desert area is east-antarctic, while West Antarctica gets somewhat more precipitation. In any case, since the loss of mass by sublimation from the snow cover of the interior is a small amount and at least partially compensated by hoarfrost deposition, practically all precipitation that falls on the surface contributes to the maintenance of the huge ice-load of the continent. So it is that as much as 75% of the total supply of fresh water on Earth exists in the form of ice, with 90% of this "total available stock" lying in Antarctica.

Also the peripheral regions have their peculiarities: First of all the great ice shelves, the Ross Ice Shelf being the largest with about $7{\times}10^5$ km^2, approximately the size of Texas, or twice the size of West and East Germany combined. Equally impressive are the ice shelves' products, the many small and occasionally a few giant tabular icebergs, up to $5{\times}10^3$ km^2 surface area.

The continuous advance of the ice shelves, with speeds averaging between 500 and 1500 m per year (Robin, 1972), can seriously threaten the meteorological stations standing on or in the snow and ice. ELLSWORTH Station on the Filchner Ice Shelf was given up in 1963, after it had moved several kilometers northward and ominous cracks had appeared to the station's south. In the first publication of IGY data (U.S. Weather Bureau, 1962) it is said: "the station is located about 1 mile (!) from the Weddell Sea and 80 miles from the nearest land. It is situated on what might be termed an ice peninsula with water to the west, north, and east." The IGY plan of the U.S. had been to put ELLSWORTH Station on firm ground at Cape Adams of Bowman Peninsula (75°S, 62.3°W) in one of the least explored corners of the cont nent. However, the ice of the south-western Weddell Sea was too strong for the cargo ship accompanying the icebreaker Staten Island. The station was then set up on the ice shelf at 41°W, only about 50 km distant from the Argentine Base BELGRANO, already two full years in operation (Schneider, 1969; Ronne, 1979). HALLEY Station, active since 1956, is getting in 1983 its fifth new housing on the Brunt Ice Shelf (personal communication, D. Limbert, BAS). More productive still was the Ross Ice Shelf, pushing not only its ice masses seaward but also the remains of LITTLE AMERICA III, Admiral Byrd's expedition headquarters of 1940 and 1941. In January 1963, the U.S. Navy Icebreaker Edisto happened to come along a small iceberg from whose top the old radio towers stood up while a sideways open room appeared on the berg's precipitous sidewall, about eight meters below surface, and 500 km westnorthwest from its original place (Am. Polar Soc., 1963).

As such bergs drift into the southern ocean and slowly break up, disperse, and continually melt, widely extended water masses are cooled noticeably, and so is the air above the waters. The effect on the weather in and around a field of bergs is felt almost immediately as fog forms and persists tenaciously (Lewis, 1975). The iceberg-cooling effect on the climate of much wider areas is potentially even more important, in particular for the South Atlantic (Schwerdtfeger, 1979; more on this cooling in Chapter 6).

In the same sense, though with more regularity, the seasonally varying extension of the pack ice[*] belt all around the continent is of great importance for weather and climate because water absorbs most of the incoming solar radiation while ice does not. This means that the surface energy budget of an open polar ocean area differs decisively from that of an ice-covered one. Fig. 1.2 shows the limit of a 9/10 or more pack ice cover in February and in September 1982. The multiannual average ice-covered areas at the end of summer and winter are 4 and 22×10^6 km^2 (Gordon and Taylor, 1975).

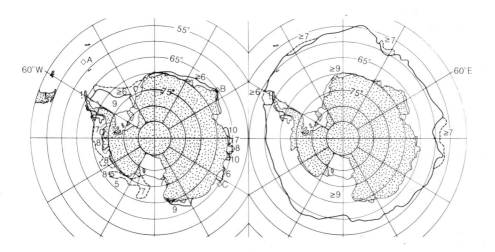

Fig. 1.2. Ice limits around Antarctica, 25 February and 16 September 1982, after the weekly charts of the Naval Polar Oceanography Center, Suitland. Size of the icebergs on the February chart: A = 65×28 km^2, B = 37×28 km^2, C = 37×15 km^2.

1.2 METEOROLOGICAL OBSERVATIONS AND INSTRUMENTATION

Here might be the place to review the ways and means by which Antarctic meteorological information has been obtained in the past, and how it is done today. The basic material consists of the hourly or three hourly observations and measurements near the surface in the conventional form of synoptic meteorology combined with upper air measurements made once or twice daily at the major stations. In modern terms one might say such a program provides the "ground truth". To maintain quality work at antarctic stations every day all

_____
[*] The term 'pack ice' is used in a wide sense to include any area of sea-ice, other than fast ice along the coasts, no matter what form it takes or how it is disposed. (WMO Sea Ice Nomenclature, 1970, T.P. 145, 147 pp.)

year round can be a hard and difficult job even with the help of modern technology. In the first decades of the century, it often has required heroic
efforts. Parts of some unforgettable stories will be quoted in Chapter 3. A
collection of classical "Antarctic Photographs 1910-1916" has recently been
published by the National Gallery of Victoria, Melbourne (1979).

For the continent-wide studies and daily analyses of synoptic meteorology and later their climatological interpretation, the number of stations and
their specific location are of greatest importance. A roster of all stations
south of 60°S with a meteorological record of at least two years has been added
in the Appendix, Table A1. Fig. 1.3 shows the location of most stations mentioned in the following chapters. Bear in mind that the old explorers could
not afford to be choosy. Sometimes they were happy to find any part of the
coast with an open inlet or a stretch of fast ice where disembarking appeared
possible and construction of safe winter quarters feasible. A bit later, not
all station locations in and around Antarctica were selected for scientific
reasons or in the spirit of international cooperation. The classical example
still is Deception Island (63°S, 61°W), where for some 20 years there were
three stations of different nationality, one a few kilometers from the other.
Then, in 1967, Nature itself stopped the needless triplication: the volcanic
activity of Deception Island revived (Baker, 1968; Baker et al., 1969; Orheim,
1970). The stations were damaged and abandoned; now, there are none.

More serious for meteorology was and is the lack of coastal stations for
a large sector. While there were more than twenty stations (operating for at
least two years) along the coast of East Antarctica between 41°W and 167°E, for
many years there was none at all from 162°W to 69°W, more than one fourth of
the total periphery. Not until 1980 a station named RUSSKAYA was installed at
74.8°S, 136.9°W.

As far as the number is concerned, here an illustrative comparison: In
recent years there were 25 meteorological stations in all-year operation on the
35×10^6 km^2 south of 60°S, compared to 340 such stations on the 9.5×10^6 km^2 of
the U.S.A., about 50 times more per unit area. (The ratio of all-year inhabi-
tants USA/Antarctica is close to 10^6). The majority of the weather stations
were established for the scientific activities of the International Geophysical
Year (IGY) 1957/58 and remained in continuous operation ever since. One of
them is AMUNDSEN SCOTT Station at the SOUTH POLE, 2835 m above mean sea level,
now also one of NOAA's (U.S. National Oceanic and Atmospheric Administration)
four 'baseline stations' monitoring "climatically important variables at locations remote from significant anthropogenic activity". Since 1977, this unit is
fittingly housed in the so-called Clean Air Facility. The measurement program
includes the continuation of the well known series of carbon dioxide data

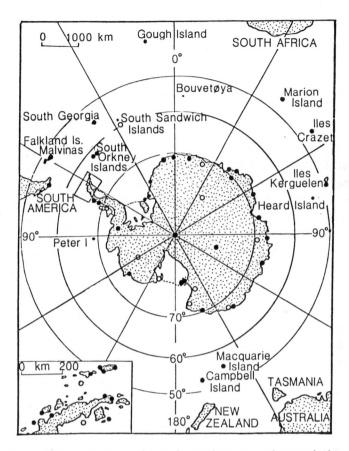

Fig. 1.3. Distribution of meteorological stations on and around the continent; the islands in the southern mid-latitudes have been included. The open circles stand for stations presently not in operation but mentioned in the text, the dark circles for stations active in 1981 and 1982 according to SCAR Bulletin (Polar Record, 1982). For the Antarctic Peninsula, completeness is not claimed. A detailed listing of the stations south of 60°S is given in Appendix A, Table A1. For most of the stations active at any time prior to 1966, useful descriptions of the place and the topography of the surroundings can be found in a report by Mather and Miller (1967).

(see Subsection 6.3.1), total ozone, surface ozone, solar radiation, turbidity, atmospheric aerosols, halocarbons, and meteorological variables as air and snow temperatures, moisture content, atmospheric pressure, wind speed and direction (Peterson and Szwarc, 1977).

One more peculiarity of the SOUTH POLE station merits to be mentioned. It is the seat of a unique and entirely voluntary organization within the Pole's winter population, the 300° Club. Membership dues: on one of the coldest days of each winter, a run from the sauna into the free outdoor atmosphere, with shoes being the only protection against the shock of a temperature change of nearly 300° Fahrenheit. (Source of information: member Mike Savage).

Of the 17 meteorological stations in the Antarctic with a continuous record of twenty or more years (prior to 1983), there are one at 90°, six between 80 and 70°, and ten between 70° and 60°S. Certainly, that is a respectable data base for general polar climatology. The same could not be said regarding the needs of synoptic meteorology, but for that part of atmospheric science a giant step ahead came a few years after the IGY: satellites in polar orbit (Fig. 4.13a). In particular over the southern ocean south of, say, 40°S, where there are few islands, few ships, and few airplanes, the new tool brought much new insight and quite a few surprises. Today, satellite information is considered absolutely indispensable for weather analysis and forecasting. No other technological achievement has helped so much to maintain an almost uninterrupted large scale surveillance of the atmosphere. By means of satellite measurements of radiative energy in several narrow wavelength bands it is now possible to determine the thickness of various atmospheric layers at many points. When that can be done with sufficient accuracy, pressure-height and winds in the free atmosphere can be computed, and the number of the upper air stations working with cumbersome balloon-radiosoundings can be reduced.

What, in the foreseeable future, probably cannot be taken over by satellites, is the determination of surface winds in or near mountainous terrain, but these winds are exactly the meteorological variable of greatest importance for any outdoor activity, short-time prediction of hazardous aircraft-landing conditions, and similar problems discussed in Chapter 3. Therefore, it is fortunate that in recent years a much less momentous technical development has successfully been carried out, the development of an automatic weather station (AWS) capable of reliable measurements even under the hard conditions of the Antarctic (Renard and Salinas, 1977; Stearns and Savage, 1981). Incidentally, the fast reporting of the AWS data to the users is done by big brother, the meteorological satellite. The AWS themselves need human attention only once a year. To help to understand the logistics of weather stations in polar regions, it should be mentioned, -- though that is seldom done -- that the continuous maintenance of a manned station including change of personnel, resupply, repair or renewal of buildings, is an expensive enterprise. Half a dozen AWS installed in the Ross Island - Ross Ice Shelf region will contribute decisively to the safety of air traffic to and from the largest antarctic city, MCMURDO, and will cost much less than one manned station. What an automatic station can and does produce is best shown in Fig. 1.4, a computer printout of the measurements of temperature, pressure, wind direction and speed during a few days in March

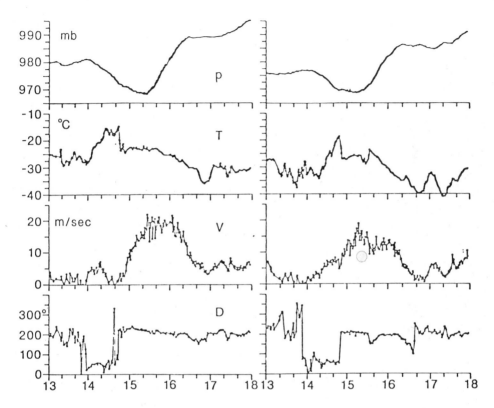

Fig. 1.4. Example of the records of pressure, temperature, wind-speed and -direction at two automatic weather stations on the Ross Ice Shelf, 13 - 18 March 1982. On the left AWS 8910, located 80 km ENE of MCMURDO. On the right AWS 8915, located 110 km SE of MCMURDO, and 110 km S of AWS 8910.

1982, by two AWS on the Ross Ice Shelf. It is obvious that such information, transmitted without time delay to the MCM Weather Office, is valuable for the aviation forecasters, and will be even more so when similar stations are deployed farther south and farther east, for instance on or near the place of the excellent IGY station LITTLE AMERICA V, 25 years ago. Successful deployment of other automatic stations by the meteorologists of the Antarctic Division of the Australian Government has been reported by Allison and Morrissy (1981).

There are other recent developments, like drifting buoys ("DRIBUs") on the waters of the southern ocean and buoys dropped from aircraft and landing with parachutes on large floes of sea ice and on selected icebergs, equally interesting for meteorologists and glaciologists (Tschernia, 1974; Ackley, 1979; Vinje, 1980.

Indeed, the drifting buoys of FGGE are a meteorological success story. Here, FGGE stands for First Global GARP Experiment, where GARP means Global Atmospheric Research Program, a large, international project. The first global experiment was scheduled for the 12 months December 1978 through November 1979. Its main elements of operation in the high southern latitudes were:

a) The World Weather Watch, meaning, the contribution of all National Weather Services (members of the World Meteorological Organization) in form of their daily synoptic surface and upper air observations.

b) Operational polar orbiting satellites and research satellites.

c) Commercial aircraft automatic flight level wind reports.

d) Oceanographic research experiments. e) Southern ocean drifting buoys.

Several hundred floating-drifting buoys, measuring pressure and temperature and reporting via satellites which determine the positions, were launched from ships, distributing them to cover the vast stationless ocean areas. Later, some "reseeding" was carried out, partly by ship, partly by aircraft. Five separate missions were flown out of Argentina, Australia, and New Zealand, strategically dropping the buoys in the most needy regions by parachute from altitudes between 150 and 5200 m. During the southern winter 1979, more than 80% of the ocean area between 20 and 65°S was within 500 km of an active buoy (Fleming et al., 1979).

To complete the enumeration of new meteorological research tools used successfully in the Antarctic, the modern acoustic sounders (Neff et al., 1976, 1977, 1978) must be mentioned. A sounding device of this kind has been deployed, for instance, at the SOUTH POLE for a detailed investigation of the extremely stable boundary-layer on the Antarctic plateau, with a temperature inversion up to 30°C in the lowest few hundred meters, but with sizeable turbulence nevertheless. Such studies can build upon the most valuable measurements made already in 1967 along a 32 m high, micrometeorologically instrumented tower erected at PLATEAU STATION (79.2°S, 40.5°E, 3625 m, Kuhn et al., 1977). Not specifically made for the polar regions, but still of great interest for the analysis of the circumpolar atmospheric vortex over the southern ocean, are the GHOST balloon flights (Global Horizontal Sounding Technique) described in Chapter 4.

1.3 POLAR DAY AND NIGHT

The Earth is closer to the Sun in the southern than in the northern summer. As a result, the polar day, defined as the time interval between the last sunrise in spring and the first sunset in fall, is shorter in the south polar regions. At 90°S latitude, where there is only one sunrise and one sunset per year, the duration is 183 × 24 hours for the polar day and 182 for the polar night; at the North Pole it is 189 and 176. For outdoor activities in the polar regions, the duration of civil twilight (sun less than 6° below the horizon) or even of astronomical twilight (less than 18°) is also of interest. Some values for different latitudes are given in Table 1.1. Detailed lists and the precise definitions can be found in the Smithsonian Meteorological Tables (List, 1958).

TABLE 1.1.

Significant dates of the duration of daylight and twilight at various latitudes. Unobstructed horizon and normal refraction conditions are assumed.

	S-POLE	85°S	80°S	75°S	70°S
Sun stays above horizon until	Mar 23	Mar 10	Feb 25	Feb 12	Jan 26
Sun remains below horizon from	Mar 23	Apr 4	Apr 19	May 5	May 25
No civil twilight from	Apr 6	Apr 20	May 5	May 27	– –
No astronom. twilight from	May 11	Jun 6	– –	– –	– –
Astronom. twilight returns at	Aug 1	Jul 4	– –	– –	– –
Civil twilight returns at	Sep 8	Aug 26	Aug 10	Jul 20	– –
Sun returns at	Sep 21	Sep 8	Aug 25	Aug 9	Jul 19
Sun stays above horizon from	Sep 21	Oct 5	Oct 18	Nov 2	Nov 19
Approx. duration of astronom. twilight at winter solstice, hours:	–	–	3	4	5
Approx. duration of civil twilight at winter solstice, hours:	–	–	–	–	2

Chapter 2

RADIATION AND TEMPERATURE CONDITIONS NEAR THE SURFACE

The main point in the first two sections of this chapter is to discuss the gains and losses of radiative energy at the surface, i.e., the irradiance = radiant flux per unit area incident upon a surface, and the radiant emittance = radiant flux per unit area emitted from the surface. The basic unit is watt m^{-2}, but for climatological considerations it is convenient to use monthly or seasonal sums expressed in MJ m^{-2}, million joules per meter square (1 MJ m^{-2} = 23.9 cal cm^{-2}). In many applications to meteorological problems one distinguishes between shortwave and longwave radiation. This is justified because only 0.4% of the total extraterrestrial solar radiant energy is transmitted in wavelengths greater than 5 micrometer, and only 0.4% of the total radiation from a body like the Earth is emitted in wavelengths smaller than 5 micrometer (Paltridge and Platt, 1976).

The last three sections call the reader's attention to characteristic temperature regimes in the boundary layer of the antarctic continent and adjacent ice shelves, regimes that are directly related to radiative processes.

Isotherm maps of Antarctica and the surrounding waters for different seasons and various height or pressure levels have not been included in this book. Not much new information could have been offered since the publication of chapter 4, vol. 14 of the World Survey of Climatology (Schwerdtfeger, 1970a) and chapter 3 of the Am. Met. Soc. Meteorological Monographs Nr. 35 (van Loon, 1972). Instead, 20 or more years series of annual temperatures, and averages of monthly temperatures are given in Appendices 2 and 3. As far as possible, data available through correspondence have been used, including the year 1982. Since the recently published "World Weather Records" give data only up to 1970, and the "Monthly Climate Data for the World" are incomplete and unchecked, the Tables in the Appendix could be useful for climatological studies.

2.1 DIFFERENT RADIATION REGIMES

In the years past, radiation measurements have been carried through at several places deep in the Antarctic, not far from the parallel 78°. Three of them, VANDA in the Wright Valley (one of the "Dry Valleys" west of the McMurdo Sound), LITTLE AMERICA in the northeastern section of the Ross Ice Shelf, and PLATEAU STATION in the highest region of East Antarctica, have been chosen to show specific radiation regimes and their relationship to the temperature conditions near the surface. The differences are large enough to make it irrelevant that the measurements at the three stations have not been made in the same years. Table 2.1 summarizes the various fluxes of radiant energy as well as

TABLE 2.1 Comparison of radiation and temperature regimes under different conditions of surface: VANDA, Dry Valleys, 77.5°S, 75 m, 1969 and 1970;
LITTLE AMERICA, snow covered Ross Ice Shelf, 78.3°S, 40 m, 1957 and
1958; PLATEAU STATION, 79.2°S, 3625 m, 1967). MJ stands for 10^6
Joules. Sources: Hoinkes, 1961; Thompson et al., 1971; Riordan,
1975; Kuhn et al., 1977.

		October to February Sun above horizon* 5 months sums	April to August Sun below horizon* 5 months sums	Year annual sum	Units
a) Global radiation G	VAN	3020	10	3160	M J m^{-2}
	LAM	2900	20	3140	" "
	PLA	4180	20	4500	" "
b) effective shortwave radiation E_S	VAN	2380 (.22)	7	2470	" "
	LAM	390 (.86)	1	420	" "
	PLA	650 (.84)	3	700	" "
in parentheses:		mean albedo			
c) effective longwave radiation E_L	VAN	-1450	-580	-2180	" "
	LAM	- 300	-220	- 610	" "
	PLA	- 620	-200	- 950	" "
d) all wave net radiation NR	VAN	930	-570	+ 290	" "
	LAM	90	-220	- 190	" "
	PLA	30	-200	- 250	" "
e) mean cloudiness	VAN	5.1	3.4	4.2	tenths
	LAM	7.7	5.7	6.9	"
	PLA	4.0	3.1	3.7	"
f) mean temperature differences	VAN - LAM	8.2	-1.1	3.8	°C
	LAM - PLA	29.7	36.2	32.6	"

* most of the time

seasonal and annual averages of temperature.   Monthly averages of radiation
values for eight stations with longer records, all in different environments,
are given in the Appendix, Table A 3.   More data of this kind have been
published by Dolgina et al. (1976).

a) The global radiation G, i.e., the sum of direct and diffuse solar
radiation incoming on a horizontal reference area, is on the high plateau much
larger than in the Dry Valleys and on the Ross Ice Shelf. The mass of air
through which the solar energy is transmitted is less than two-thirds of the
mass above the other two places (the annual mean surface pressure at PLATEAU
STATION is about 610 mb).   Kuhn et al. (1977, p. 49) introduced the ratio of
measured global radiation to the extraterrestrial energy flux as a useful value
describing, for any place, independent of latitude, the effect of the atmosphere on global radiation. This ratio (average for the daylight period)
amounts to 0.80 on the high plateau, values between 0.55 and 0.65 at places

14

near sea level. VANDA is one of the few stations where there are mountains, here mostly the Olympus Range to the north, so close that the direct solar energy flux reaching the surface is reduced to some extent. If that were not so, the difference between VANDA and LITTLE AMERICA would be larger because at the latter the frequency of occurrence of opaque low clouds, capable of reflecting up to 50% of the incoming energy, is much greater. This, in turn, is due to the fact that the advection of relatively warm, moist air from the north-quadrant in the lower one-third of the troposphere is obstructed at VANDA, but not on the ice shelf.

b) One of the main factors in any analysis of the shortwave fluxes is the reflectivity or albedo of the receiving surface ($\underline{a}$ = ratio of reflected over global radiation). It is small in the "oasis" regions which remain free of snow in most or all of the sunlit time; moderate in the case of rough ice and irregular, rocky surfaces with thin or broken snow cover; and decisively large on the great snow covered ice shelves and vast snow fields of the interior. Monthly mean albedo values for various stations are shown in Fig. 2.1. Table 2.1 gives the five-months and annual sums of the effective shortwave radiation, the global plus the (negative) reflected shortwave flux, $E_S = G(1-\underline{a})$.

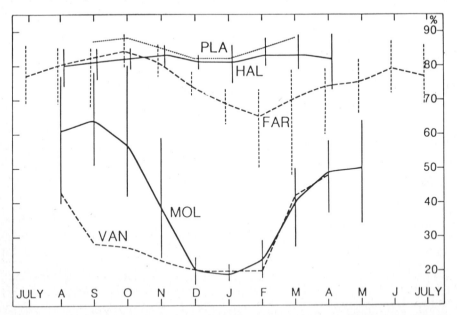

Fig. 2.1. Monthly mean values of the albedo (%) in various climate regimes. For HALLEY (HAL), FARADAY (FAR), and MOLODEZNAYA (MOL) the values are ten year averages 1963 - 1972; the end points of the vertical lines indicate the largest and the smallest monthly values of each station. Data from Farman and Hamilton (1978) and Dolgina et al. (1976). PLATEAU Station (PLA) for 1966 and 1967, from Kuhn et al. (1977); VANDA (VAN) 1969 from Riordan (1975).

c) The effective longwave radiation E_L is the sum of the upward directed (negative) terrestrial radiation R_L↑ and the downward directed (positive) atmospheric radiation (or counter-radiation) R_L↓. For many meteorological applications, the solid or liquid surface from which R↑ is emitted, and the atmosphere as source of R↓, are treated as <u>gray</u> bodies. That means equal emission (and absorption) at every wavelength in the infrared part of the spectrum is assumed. Consequently, the Stefan-Boltzmann law $R = \varepsilon\sigma\, T^4$ can be used (T = temperature in °K, σ=St.-B. constant = 5.67×10^{-8} W m^{-2} K^{-4}, and ε = emissivity). Then it is relatively simple to determine the outgoing radiation R↑, since ε varies little (Table 2.2) and the temperature of the surface can be measured. Note that the difference between the real surface temperature T_s and the air temperature T measured 1.5 to 2 m above the surface and loosely called the "surface temperature" is not always negligible (Table 2.3). An easy way to estimate the contribution of the atmospheric radiation to the longwave energy budget is to introduce a dimensionless parameter, the so-called Ångström ratio $Å = E_L/R↑ = (R↑ + R↓)/R↑$. This ratio characterizes the longwave radiation conditions at a given time in an area of uniform surface. A small value of Å indicates that the downward directed energy flux counterbalances a large fraction of the flux going upward from the surface.

TABLE 2.2 Some longwave emissivity values

a)

Water	0.92 - 0.96	coarse gravel	0.90 - 0.92
Ice	0.95 - 0.97	moist, bare ground	0.95 - 0.98
Snow	0.87 - 0.95*	frozen soil	0.93 - 0.95

b)**Dense stratus or stratocumulus clouds,   > 0.9
     water clouds, > 500 m vertical extent

     Ice clouds   0.1 - 0.5
     (Kuhn and Weickmann, 1969; Murcray et al., 1974).

_____

* the lower value in cold, small grained snow,
   the higher one in melting snow (Mellor, 1977)

**these values are estimates for polar regions.

It turns out that on the antarctic plateau, represented by SOUTH POLE, PLATEAU STATION, VOSTOK AND MIZUHO, Å is comparatively large (about 0.30) in the short summer, going back to values less than 0.20 in the long winter. This is mostly due to the low surface temperatures of the winter leading to small values of R↑, and the presence of an 20 to 25° warmer and moister layer at 500 to 1000 m above surface leading to larger values of R↓. In the coastal regions, with LITTLE AMERICA, HALLEY, MIRNY, FARADAY and others, the summer values of Å are around 0.15, in the winter between 0.10 and 0.15. There is a strong

TABLE 2.3 Monthly mean difference between surface and air temperature.
Molodezhnaya 1963-67 (Averyanov, 1975); Mirny, Pionerskaya and Vostok 1956-58
(Rusin, 1961); SOUTH POLE 1958 (Dalrymple, 1961, 1966). °C. The minus sign
means colder at the ground.

	MOL	MIR	PIO	VOS	SOUTH POLE	2835 m
	both near the coast		2740 m	3488 m	0 to 1 m	0 to 2 m
JAN	7.2	0.1	0.2	1.0	-	-
FEB	4.0	0.0	0.0	-0.1	0.2	0.2
MAR	0.7	-0.7	-0.7	-1.8	-0 7	-0.8
APR	-1.0	-1.1	-0.8	-1.4	-0.9	-1.0
MAY	-1.0	-1.2	-1.0	-2.2	-1.8	-1.9
JUN	-0.9	-1.3	-0.8	-2.5	-1.9	-2.0
JUL	-1.4	-1.1	-0.9	-1.7	-1 1	-1.2
AUG	-1.4	-1.3	-1.0	-1.7	-1.4	-1.5
SEP	-0.8	-1.0	-0.7	-2.2	-2.4	-2.6
OCT	1.3	-0.4	-0.7	-0.6	-0.1	-0.2
NOV	4.2	-0.2	0.0	-0.2	1.4	1.4
DEC	6.5	-0.1	0.3	1.6	-	-

contrast between VANDA and LITTLE AMERICA, corresponding to the differences in
cloudiness (Table 2.1). In fact, the Å values for VANDA appear to be larger in
summer and winter than on the plateau, but more data will be needed to ascertain such a finding. Altogether, accurate measurements of the longwave radiation fluxes have been, and often still are, very difficult to obtain as long as
the Sun is above the horizon. In the words of Kuhn et al. (1977, p. 63), it is
"the purgatory of longwave radiometry". All the values above quoted are seasonal averages, December and January going as summer, April to August as winter.
Variations in shorter time intervals are discussed in Section 2.2.3.

d) The sum of effective shortwave and longwave radiation, called the net
radiation NR, radiation balance, or radiation budget, is given in part d) of
Table 2.1 for each of the three stations. In the five months October to
February, the energy flux available for surface heating, melting and evaporating is at VANDA about ten times as large as at LITTLE AMERICA. The effect of
the sensible heating can be seen in the large temperature differences between
the two stations shown in the lower part of Table 2.1. In the entire Wright
Valley the processes of latent heating, i.e., melting and evaporation, are not
negligible either. A narrow marginal zone of the otherwise perennially frozen
Lake Vanda (about 6×1.5 km^2, west of the station) opens in the two summer
months. The Onyx River, rather a summertime rivulet, carries meltwater from the
glaciers at the eastern end of the valley toward the lake, a distance of about
20 km; and the dryness of the air, prevailing in the October to February period,
favors evaporation. Nothing of comparable magnitude happens at LITTLE AMERICA
where the albedo remains high (> 0.80) through all of the summer. The same is
true, of course, for the high plateau, though a slight decrease of the albedo

with increasing height-angle of the Sun exists. In the winter, the large difference between the net radiation values of the two low-elevation stations again shows the importance of cloud conditions for the atmospheric radiation, and the similar difference between VANDA and PLATEAU station can be understood as due to the very low surface temperatures and the correspondingly small values of the upward flux at the latter place.

The difference between the real surface temperature T_s, to which the radiation laws implicitly refer, and the "surface" air temperature T measured about 1.5 m above the surface and generally used in climatology and synoptic meteorology, is in the case of a thick snow cover small enough (< 2°C) to be disregarded in a qualitative interpretation of radiation values as given above. A detailed discussion of the differences between the mean annual temperatures at the height of the shelter, at the surface itself, and at 10 m below the surface has been given by Loewe (1970). The temperature difference between shelter and surface becomes really large for barren soil or rock surfaces in the summer months. The five-year monthly mean values of this difference for MOLODEZNAYA (Averyanov, 1975) in comparison to four stations with perennial snow cover (Table 2.3) show what is meant. There can be no doubt that on a sunny summer day the boundary layer receives a considerable amount of heat when and where the ground is bare and the albedo small (Fig. 2.1, MOL and VAN); the surface temperature increases, and some convection is initiated and maintained with superadiabatic lapse rates in the lowest layers. Consequently, there will also be differential heating, producing a pronounced horizontal temperature gradient between the bare ground area and neighboring or surrounding snow fields. On a larger scale, the same must be the case between ice-free coastal waters and snow covered land, fast ice or shelf ice. Under such conditions, thermal winds (Section 3.1.2) can persist in the boundary layer which modify the surface wind pattern, as will be discussed in the next chapter.

In this context it must be mentioned that the net radiative energy flux is only one, though the most important, term in the equation that describes the complete energy budget for a place at the surface of the Earth. Other contributing processes are: i. The eddy transport of sensible heat in the vertical, correlated to the stability of the vertical structure of the boundary layer, the strength of the wind, and its vertical shear. That means, for instance, that the temperatures on the interior of the continent in the eight or more winter months would be lower, and the strength of the surface inversion still greater, without this heat supply. ii. The heat conduction in snow, ice, or rocky ground, upward to the surface in the colder part of the year, downward in the short summer. iii. The latent heat flux, positive when hoarfrost forms at the surface, negative when energy is supplied for the sublimation (evaporation) process. (Liljequist, 1956; Loewe, 1957; Rusin, 1961; Dalrymple et al., 1966; Carroll et al., 1981).

These three constituents of the surface energy budget will not be treated in detail in the following chapters. This is not to say that these processes be negligible, -- they are important, -- but they belong justly to the field of micrometeorology, and contribute little to the vagaries of weather and the possible fluctuations of climate of the Antarctic, the main objects of this book.

2.2 YEAR-TO-YEAR AND SHORT TIME VARIATIONS OF RADIATIVE ENERGY FLUXES
2.2.1 Variations of Global Radiation

The radiation values given in Table 2.1 do not permit any conclusions about the possible variations of monthly or seasonal sums of the energy fluxes from one year to another. Fortunately, there are some antarctic stations of which continuous records for ten or more years have been published in recent years, among them the British stations HALLEY and FARADAY, formerly called Halley Bay and Argentine Islands, (Farman and Hamilton, 1978), and several Russian stations (Dolgina et al., 1976). HALLEY lies on the Brunt ice shelf, southeastern Weddell Sea, a location somewhat similar to that of LITTLE AMERICA. FARADAY is situated on a small island about 5 km off the west coast of the Antarctic Peninsula; its climate is milder and moister than that of any other coastal station south of 65°S. VOSTOK represents the High Plateau, and MIRNY the coast of East Antarctica. Table 2.4 shows for the month of December that the variability of G is much smaller at the higher than at the lower latitude stations, and smallest on the plateau; as to be expected, a rather close relationship exists between G and S, the latter being the duration of sunshine (more precisely the absence of opaque cloud decks, but not necessarily the often observed thin cirrus clouds). It will be seen that the variations of the monthly and seasonal sums of the other radiation fluxes discussed in Section 2.1 are considerably larger, for good reasons.

The day-to-day variations of the global radiation G are of interest because they indicate changes of the shortwave transmissivity which, in turn, depend on the varying weather conditions in the wider environment of the measuring stations. For instance, a high reaching invasion of relatively warm, maritime air from the southern ocean on the front side of a depression or a low pressure trough brings more than just a rise (above the boundary layer) of the temperatures; there will be an increase of the air's H_2O content, possibly in all three phases, and a change of the aerosol content.

Only a few, and short, series of daily and hourly radiation values have been published, most recently for MIZUHO station in the JARE Data Reports for

TABLE 2.4 Totals of global radiation G and duration of sunshine S (in hours and percentage of the possible) in the month of <u>December</u>. σ = standard deviation.   Sources: Farman and Hamilton (1978); courtesy D.W.S. Limbert, BAS; Dolgina et al. (1976).

HALLEY, 75.5°S, 35 m   FARADAY, 65.3°S, 11 m

	Years	MJ m^{-2}	hours	%		Years	MJ m^{-2}	hours	%
Average	1963-72	885	301	40		1963-72	654	150	25
σ	1963-72	24	58	8		1963-72	93	72	12
Max. of G	1965	912	347	47		1970	792	302	50
Min. of G	1963	843	243	33		1963	504	44	7

VOSTOK, 78.5°S, 3488 m   MIRNY, 66.6°S, 40 m

	Years	MJ m^{-2}	hours	%		Years	MJ m^{-2}	hours	%
Average	1958-73*	1235	725	97		1956-73*	967	304	41
σ	1958-73	49	13	1		1956-73	62	83	12
Max. of G	1968	1348	739	99		1968	1072	393	54
Min. of G	1971	1143	689	93		1970	825	192	26

*1960 and 1962 missing   *1960 missing

1979 and 1980. Unpublished data of USSR stations have been discussed by Marshunova (1980). Table 2.5 presents some information on global radiation at other stations for the month of December. Again it becomes evident that the variability of the incoming solar radiation on the high interior, here represented by PLATEAU, S-POLE and MIZUHO Station, is much smaller than at all other places listed, including BYRD where heavy cloudiness of synoptic disturbances, as described above, can easily appear. Typically, the station situated closest to the trajectories of subpolar storms has the most pronounced variability.

TABLE 2.5 Daily values of global radiation in the month of <u>December</u>. G = average of the daily values, in MJ m^{-2} day^{-1}; RMS ΔG = root-mean-squares of the day-to-day variations ΔG.

Station	Elev.	Lat.	Year	days	$\overline{G}$	RMS ΔG	Extremes	of ΔG
PLATEAU	3625 m	79.3°S	1967	31	39.2	0.8	+ 1.9	- 2.1
S-POLE	2835	90	1958-65	248	40.0	2.9	+ 7.5	- 8.8
MIZUHO	2230	70.7	1979-80	62	35.1	2.8	+ 8.0	- 9.0
BYRD	1530	80.0	58	31	31.5	5.2	+14.2	-13.1
VANDA	75	77.5	71	31	25.8	7.2	+13.8	-12.8
LI. AM.	40	78.3	57	31	25.9	5.6	+ 8.4	-15.5
ELLSW.	42	77.7	1957-58	62	30.2	6.0	+15.6	-12.8
HALLETT	5	72.3	58	31	31.2	6.1	+12.2	-16.0
WILKES	12	66.3	58	31	26.4	9.4	+15.7	-20.1

More detailed information for PLATEAU station has been elaborated by Kuhn et al. (1977). At the station's elevation the daily total of G can reach about 90% of the value valid for the top of the atmosphere. Of course, that happens only under optimum conditions, possibly with some favorably located reflecting ice clouds. The maximum daily sum recorded near the summer solstice is close to 42 MJ m^{-2}, equal to the often debated "record" of 1000 cal cm^{-2} (Rusin, 1961). The maximum at the SOUTH POLE, measured in December 1958, is 41 MJ m^{-2}. Under bad weather conditions with snowfall or drifting snow the G values decrease to about 60%, in extreme cases to 40%, of the extraterrestrial radiation (Kuhn et al., 1977, Fig. 6 and Table 5). This refers to the High Plateau where the total H_2O content of the air in a vertical column through the atmosphere never reaches 1 gram cm^{-2}. At places near sea level inclement weather leads to much stronger flux depletion, and G can vary strikingly from day to day.

2.2.2 Variations of albedo and effective shortwave radiation

The interannual variations of the monthly or seasonal sums of the effective shortwave radiation E_S can be large in regions where the albedo a changes substantially and not necessarily every year at the same time or same rate. The wide range of monthly mean a-values in spring and fall at FARADAY and MOLO (Fig. 2.1) shows what is meant. In some segments of the coastal regions, like the west side of the Antarctic Peninsula and parts of the coast of Enderby Land, the winterly snowcover can begin to age, partially break up, and melt as early as the weeks prior to the summer solstice. Then the albedo decreases at a time when the global radiation incrases. Consequently, the effective shortwave radiation [$E_S = (1-a)G$] increases sharply. However, when at that time of the year fresh snow has fallen and cold winds from the interior keep the daily maximum temperature well below freezing, a remains large and the increase of E_S is minor. On the other hand, in late summer G decreases, and in some years and at some places, though not always nor everywhere, the snowcover has given way to the barren ground, which means that the albedo has decreased strongly. Then the effective radiation, the energy that heats the ground, diminishes but little. All this tells us that the chances for large year-to-year variations of E_S are very good; they depend on past as well as present weather conditions.

For stations at higher latitude or elevation, on the ice shelves or the plateau, the variations of the monthly means of albedo a are smaller. For the magnitude of E_S, however, the factor (1-a) is decisive; since the albedo a is slightly more than 0.80 most of the time, a small variation leads to a relatively large change of (1-a), and hence of E_S. There is plenty of evidence (Rusin, 1961; Dolgina et al., 1976; Kuhn et al., 1977; Carroll and Fitch, 1981) to show that the albedo diminishes by a few percent when the Sun's elevation angle increases. This is confirmed by mean monthly values (HAL and PLA in Fig. 2.1) as well as by the analysis of hourly data of global and reflected radiation on the high plateau. Still, temporary complications are always possible, due to large snowdrifts upwind of the station, formation of sastrugi, varying cloud conditions, and others. A comprehensive discussion of the optical properties of snow was recently given by Warren (1982).

As far as the day-to-day variations of E_S are concerned, rapid changes of G, as discussed above, contribute more than changes of the albedo. Naturally, the latter is not true when a copious snowfall comes down on previously barren ground, raising the albedo from, say, 20% to 80%, as it can happen in the oasis regions.

Finally, the possible effect of large scale dust deposition on the white antarctic snow fields should be mentioned. Such deposition could be due to natural or man-made events, as volcanic eruptions or industrial development on the continent. A decrease of the albedo from the present 85% to, say, 60% would bring about a highly significant change of the energy budget for the affected area (Lettau, 1977). The climatic consequences would depend, of course, on the duration of the modified surface conditions while Nature again and again might work to restore the original ones.

In this context it may be noted that the term albedo by virtue of its definition embraces a wider range of wavelengths than the visible part of the electromagnetic spectrum alone. The flux of solar energy in the near infrared amounts to somewhat more than one-third of the total flux. Therefore, the dependence of the albedo of different kinds of surface upon the wavelength of the incoming radiation cannot be disregarded (Fig. 2.2, courtesy of M. Kuhn).

2.2.3 Variations of Ångström ratio and effective longwave radiation

For an understanding and estimate of the variability of E_L, it is convenient to make use of the Ångström ratio $A = (R_L\uparrow + R_L\downarrow)/R_L\uparrow = E_L/R_L\uparrow$, which serves to characterize the longwave radiation conditions at a given place and time (Section 2.1). One must bear in mind that the interannual variations of monthly and seasonal temperatures in the high southern latitudes are rather small, as Table 2.6 indicates. Since a rise of the surface temperature by 1°K leads to an increase of $R_L\uparrow$ by less than 2%, a pronounced change of $E_L = A \times R_L\uparrow$ cannot be brought about without a substantial change of the Ångström ratio A. An example taken from the ten year record of radiation measurements at HALLEY will illustrate this statement:

May   1963: $\overline{T}$ = - 26.1°C, E_L = - 35 MJ m^{-2} , A = 0.06 approx.
July 1964:   - 29.9   - 85   0.16 approx.

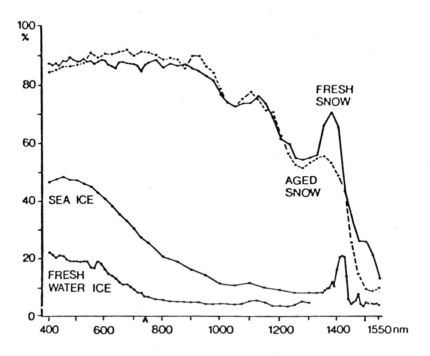

Fig. 2.2. Albedo as function of wavelength for different types of snow and ice (Kuhn and Siogas, 1978). The upper limit of the visible range of the spectrum is marked at 740 nanometers = 0.74 micrometers.

Table 2.6 Average and standard deviation of the monthly mean temperatures near the surface, and root-mean-square of the interannual variation of the monthly means, RMS ΔT; July and December, for four stations with long series of data.

	SOUTH POLE		HALLEY (75.5°S)		MIRNY (66.6°S)		FARADAY (65.3°S)	
	July	Dec	July	Dec	July	Dec	July	Dec
T̄	-60.1	-27.7	-29.2	-5.3	-16.6	-2.4	-10.7	-0.4°C
σT	2.4	1.7	2.8	1.1	2.6	0.9	4.5	0.5°C
RMS ΔT	3.6	2.5	3.3	1.5	3.8	1.3	5.0	0.6°C
Record	1957-82		1956-82		1956-82		1947-82	

Obviously, $R_L\uparrow$ is here not the decisive factor; what is going on in the troposphere, and mostly in its lower layers, is more important for the variability of E_L. The latter also applies to the summer values of E_L. The HALLEY data suggest Å between 0.25 and 0.35. At PLATEAU station Kuhn et al. (1977) found $\text{Å} = 0.29$ in November, $\text{Å} = 0.37$ in December 1967. Much wider variations of Å, even down to negative values, appear only in shorter time intervals, like hours or, at most, exceptional days at any time of the year. They are directly related to the weather, and should therefore be given attention. Unfortunately, however, very few series of hourly radiation data of Antarctic stations have appeared in the annals or station records.

Hourly values of $R_L\uparrow$ and $R_L\downarrow$ at MIZUHO in the winter 1979 and 1980 (Yamanouchi et al., 1981; Ishikawa et al., 1982) indicate that in 89% of all hours May through August the Ångström ratio Å was positive, 6% zero, 5% negative. For all 246 days, the frequency distribution (in %) of the Å values is the following:

Å	-0.01 to 0.00	0.01 to 0.10	0.11 to 0.20	0.21 to 0.30	0.31 to 0.36
days	3	22	28	36	11 % .

The overall average was 0.18, $\sigma = 0.09$.

How much the occasional rapid changes of weather conditions can influence the effective longwave radiation E_L, becomes evident in an analysis of measurements obtained in the winter 1965 at the S-POLE by a CSIRO (Australia) net radiometer (Schwerdtfeger, 1968). Table 2.7 shows E_L as a function of time for seven typical cases of change from overcast to clear sky, synchronized so that hour zero is the last time the cloud deck was observed. The drastic change of all three parameters, E_L, T and Å, does not require further comment. Cloud observations in the winter 1965 were taken at the SOUTH POLE only four time per day. Under such circumstances, a more exact time of the clearing of the sky could be found from continuous net-radiation records than from the synoptic observations.

It must also be borne in mind that the quality of cloud observations during the polar winter night depends on the presence of moonlight. Nevertheless, experienced observers can well distinguish a relatively dense low cloud deck from the less dense clouds generally classified as cirrostratus. Estimates of the height of the base of the denser clouds, occasionally supported by timing the disappearance of a little light attached to the radiosondes, are between 300 and 800 m above the surface, that is in the warmest layer of the troposphere, in the upper part of the inversion. From all six-hourly cloud observations and the simultaneous measurements of E_L carried out in the winter half-year of 1965 at the SOUTH POLE, frequencies and average values, respectively, are listed in Table 2.8. The occurrence of eight cases of positive E_L

TABLE 2.7 Variation of the net radiation E_L (here = NR), surface temperature T, and Angström ratio Å, as a function of cloudiness. Average of seven synchronized cases of rapid clearing at the SOUTH POLE, winter 1965. Time 0 indicates the last (six-hourly) observation of low-cloud overcast. BS = blowing snow.

Time hours	Cloudiness low	high	E_L kJ m^{-2} h^{-1}	Å	-T °C
-12	9	x	9.6	.02	51
			0.4	-.01	51
- 6	10	x	22.6	-.04	47
			38.1	-.07	46
0	9	x	11.3	-.02	44
			-49.8	.09	47
6	0	0	-74.1	.15	50
			-76.2	.16	53
12	0	0	-90.4	.20	56
			-76.6	.17	57
18	0	0	-80.0	.18	58
			-73.7	.17	59
24	BS	x	-70.8	.17	61
			-60.7	.15	62
30	0	10	-55.7	.14	62

TABLE 2.8 Longwave radiation flux E_L (here = net radiation NR) as function of cloudiness, at the SOUTH POLE.

Sky conditions	10/10 As*	10/10 Cs	Clear Sky
Number of observations	82	55	412
Average E_L	+ 26	- 19	- 70 k J m^{-2}hour^{-1}
Number of cases with E_L > 0 (downward)	63	18	8

---

*Unfortunately, since the inauguration of the AMUNDSEN-SCOTT Station at the SOUTH POLE in 1957 the meteorological observers have given to the low clouds the inappropriate name Altostratus (As), thus referring to the height of the cloud base above sea level, instead the height above surface as the international code rightly prescribes.

with clear skies could be understood as due to measurements made only a short time before the clearing.

As far as only the amount of clouds and the temperature near the surface are concerned, statistics for eleven winter half-years at VOSTOK are given in Table 2.9.

TABLE 2.9 Air temperature near the surface as function of cloudiness, at VOSTOK, 3488 m. Averages of six-hourly synoptic observations in the winter months April to September, 1958-61 and 63-69.

Cloudiness	mean temperature	number of observations
< 2/10	- 68.6 °C	3587
3 to 7/10	- 66.0	1550
> 8/10	- 60.9	1451
all obs.	- 66.3	6588

2.2.4 Variations of net radiation

Complete, published series of monthly values of net radiation NR for ten or more years exist for several places. Farman and Hamilton's (1978) work includes a detailed description of the instruments used and the problems encountered at HALLEY and FARADAY. For these two stations and for MIRNY (with a longer uninterrupted record than the others) the standard deviation σ_{NR} has been added in Table A 3 R as a measure of the year-to-year variability of the monthly and annual sums of NR. The data in the Table show how the radiation regime of the oasis-type places, MOLODEZ, NOVALAZ and VANDA, differs decisively from the regimes at other stations, at lower and higher latitude. It must be borne in mind, though, that these three represent only a very small fraction of the total surface of Antarctica. The last two stations in the Table, VOSTOK and MIZUHO, describe the conditions on the plateau. Their radiation values are remarkably similar one to another, despite of eight degrees difference in latitude and 1200 m in elevation.

There is little information available on the periodic variation of NR in the course of a summer day in the climatically different regions of the Antarctic. The situation is simple only at the SOUTH POLE itself where the change of zenith distance of the Sun within 24 hours, as well as any diurnal variation of cloudiness, are negligible. There remain only aperiodic variations of cloud cover and cold or warm advection; both processes are most efficient when occurring in the warmest layer (see Sections 2.3 and 2.4), a few hundred meters above surface.

The diurnal variation of NR at lower latitude, but still on the antarctic plateau, is shown in Fig. 2.3 for MIZUHO. It makes sense that the rapid change of NR in the afternoon of the clear sky days is due to the large $R_L\uparrow$ value when the surface temperature is relatively high, combined with a small $R_L\downarrow$ value from a cloudless sky. More measurements with modern instruments are needed, however, to ascertain further interpretation.

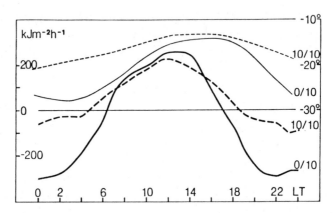

Fig. 2.3. Diurnal variation of net radiation (NR) and surface temperatures at MIZUHO on January days with clear and with overcast sky, and with moderate winds. Clear sky data are in solid lines, overcast in dashed lines. Temperatures in thin lines and right-hand side scale, NR in heavier lines and scale to the left. The data are for 6, 7, 8, and 1, 3, 10, January 1980, respectively; all from JARE Data Reports, No. 73.

2.3 THE POINTED SUMMER AND THE CORELESS WINTER ON THE HIGH PLATEAU

The peculiar annual march of temperature near the surface of the antarctic plateau is essentially coined by the radiation conditions, in winter as well as in summer. The resulting regime is not only of theoretical and climatological interest, but also had, only a few decades ago, unforeseen consequences for antarctic explorers. Therefore, a bit of polar history might be in place (Shackleton, 1911; Amundsen, 1912; Barrie, 1913; Siple, 1959).

When in the first months of 1910 Amundsen in Norway and Scott in England prepared their expeditions to the South Pole, some information regarding the conditions to be expected on the Antarctic Plateau was already available. In one of the greatest pioneering journeys of all time, Shackleton and his companions had discovered and mastered the Beardmore Glacier in December 1908 and had progressed south to latitude 88°23'S, only 180 kilometers from the Pole. They had to turn back that close to their goal since their supplies, food and fuel were insufficient, and they made it back to their base on Ross Island only due to the stamina of their leader. For 15 days, 1-15 January, they had stayed south of 87°S at an elevation of approximately 3,000 meters above sea level. Shackleton gave a clear description of the Antarctic Plateau, including the daily temperatures which fluctuated between -23° and -40°, and averaged -29°C. He could not get any notion of the duration of the south polar summer, and neither could Amundsen who reached and left the Pole in December 1911 and was down again on the Ross Ice Shelf on 3 January 1912. The observations of Scott's party indicate a pronounced decrease of temperature in the second half of January. Without observations of other years, however, that also could be interpreted as a singularity of the disastrous year 1912. Today, ample information is available; the AMUNDSEN-SCOTT (South Pole) station is in operation since January 1957. Its impressive record reveals two remarkable features of the temperature regime. First, the south polar "summer" is very short; not more than about 30 days between mid-December and mid-January deserve that name. Two weeks later, the average temperature already lies about 8°C below that of the warmest days. This early and fast cooling on the continent's high plateau was one of the adversities contributing to the tragic end of Scott's journey.

The extreme shortness of the polar summer is characteristic for a wide area. For the station VOSTOK, at 78.5°S, Averyanov (1972) published the eleven year averages of temperature per decade of the four months November-February; the minus sign applies to every value.

November			December			January			February		
1	2	3	1	2	3	1	2	3	1	2	3
47.0	43.3	39.9	35.5	32.0	31.0	30.8	32.6	35.0	39.8	45.0	49.0°C

In some of the climatological statistics for that station, the year is divided into four seasons: Summer = December and January, fall = February and March, winter = April through September, and spring = October and November.

This shortness of the summer is caused by the variations of global radiation G and the albedo. On the average of a number of years, the maximum of G has to be expected at the summer solstice. Already in the preceding weeks the albedo decreases, due to increasing solar elevation and a slight metamorphosis of the snow cover (Kuhn et al., 1977; Carroll and Fitch, 1981). Thus, there are two reasons why the effective shortwave radiation E_S increases or, in other words, more radiative energy can be absorbed at the snow surface. After solstice, G decreases, and the first light snowfall or inflow of drifting snow restores surface conditions favoring larger albedo values. Hence, there are two reasons to expect a fast decrease of E_S in January. It would be a mistake, of course, to envision for every summer exactly the same variation of the surface temperature, because cloudiness and surface winds can modify the picture. Nevertheless, the trend is evident, and the "pointed summer" the most probable temperature pattern (A.J.U.S. 1977, 12(4): 156-159).

The second peculiar feature of the annual march of the temperature near the surface is the better known "coreless" winter. (This term, in German "kernlos", was coined by J.v. Hann (1909) referring to the annual march of temperature at early stations in the Antarctic.) Again, a short historical note will be added, this time with a happy ending.

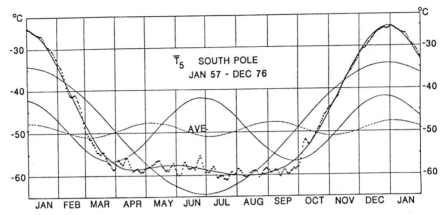

Fig. 2.4 SOUTH POLE temperature regime. Explanation of lines in the text.

When Paul Siple came to the South Pole on 30 November 1956, the first scientist to stay for a full year at that far-out place, he did not hesitate to dig, in 4 days of hard work, a 5.5 meter-deep pit. The purpose was to measure the temperature which at that depth comes close to the mean annual value. Knowing about the summer temperatures on the plateau from Amundsen's and Scott's reports, and assuming the temperature of the coldest month should be as much below the annual mean as the summer values are above it ("like it is in most other places"), he concluded that the average temperature of the coldest month might drop below -84°C (-120°F), a possibility he considered "half in apprehension and half in excitement" (Siple, 1959). Ten months later, he was surprised as well as relieved to find that the winter of the Antarctic Plateau is different. In 1957, the coldest month was September with a mean temperature of -62.2°C and the lowest minimum of -74°C. Looking at the averages as shown in Fig. 2.4, one sees that already in the last days of March it is nearly as cold as in the six following months. In the 26-year average 1957-82, July is the coldest month, as Paul Siple expected, but only by two tenths of a degree; in these 26 years April was twice and September seven times the coldest month.

Again it becomes evident that the temperature regime near the surface of the Antarctic Plateau is essentially determined by the radiative energy budget. During the winter half year, the temperature inversion in the boundary layer is very strong. At the SOUTH POLE, for instance, the temperature increases from about -58° at the surface to -36° in the isothermal layer 500 to 1,000 meters above, on the average. This means the energy flux outgoing from the surface, $R_L\uparrow$, is comparatively small and the atmospheric radiation $R_L\downarrow$ from the warmer and moister layer large. It must be considered, of course, that the emissivity ε of a snow surface is considerably greater than that of the warmest atmospheric layer. Still, the Ångström ratio on the high plateau, about 0.30 in the summer, decreases to less than 0.20 in winter. An additional factor is the eddy flux of sensible heat downward (Dalrymple et al., 1966; Schwerdtfeger, 1970a,b; Miller, 1974; Lettau, 1977).

The remarkable temperature increase from the low values in the first days of June to the winter maximum later the same month, as apparent in Fig. 2.4, calls for a comment. Statistical analysis indicates that this feature can well be due to chance. Only if it were to appear with appreciable magnitude in the record of many more years would one be justified in accepting it as a real phenomenon whose cause should be investigated. The same applies, of course, to the less pronounced changes between March and early October, and to an apparent rhythm of approximately 30 days which an imaginative viewer may find in the line of dots in the figure.

Fig. 2.4 and Table 2.10 give more information on the coreless winter and the entire, average seasonal variation of temperature on the Antarctic Plateau. In the graph, the heavy dots represent the five-day mean values of the 20-year average of the daily mean temperature for every calendar day, plotted for the central day of each five-day period. Hence, every dot represents 5 × 20 = 100 days, smoothing out the random variations more efficiently than a simple 20 year average for each calendar day can do. The thin dashed, solid and dash-dotted curves show the first, second and third harmonic component of the annual march, and the solid line close to the daily dots represents the sum of the three components. Table 2.10 contains the results of the harmonic analysis also for the two other stations on the plateau with a sufficiently long record.

TABLE 2.10. Results of harmonic analysis of the annual march of temperature at three inner-antarctic stations.

Parameters	SOUTH POLE 1957-80	VOSTOK 1958-80*	BYRD 1957-69
Date of max.	1 Jan.	3 Jan.	14 Jan.
Amplitude	15.2	17.0	10.6°C
% of total variance	81	85	89%
Dates of max.	3 Jan., 2 Jul.	4 Jan., 3 Jul.	2 Jan., 1 Jul.
Amplitude	7.0	7.1	3.5°C
% of total variance	18	15	10%
Amplitude	1.5	0.7	0.3°C
% of total variance	< 1	< 0.2	< 1%

*no data for 1962

Summarizing, it can be said that the average temperature conditions and seasonal variations at or near the surface of the Plateau are well understood. In addition, a few remarks on extreme values might be in place. Everywhere in the interior of the continent the lowest winter temperatures are mainly the result of enduring negative values of the effective longwave radiation combined with low windspeeds. This has led to extreme minimum temperatures of -80.6° at the SOUTH POLE, -89.5° at VOSTOK. In contrast, an analogous statement regarding extreme maximum summer temperatures, as declaring a large positive radiation budget the main cause, would go too far. Advection of exceptionally warm, maritime air by strong winds from lower latitudes is another important source that eventually can bring higher temperatures than a large positive NR with weak winds. Highest values of -13.6° at SOUTH POLE, -15.7° at VOSTOK have been observed (Anonymous, 1977; Sinclair, 1981).

2.4 THE GREAT TEMPERATURE INVERSION OF THE LOWER TROPOSPHERE

A general picture of this continent-wide phenomenon is given in Fig. 2.5, computed for the period June to August. Nevertheless, in the highest latitudes it is approximately valid for eight, in the coastal regions four winter months. Lapse conditions (decrease of temperature with height) prevail in the short summer, with the exception of the high plateau where the monthly averages still indicate a weak inversion (Tab. 2.11).

The periodic diurnal variation of the temperature of the air near the surface in the sunlit and partly sunlit months can be quite large, as shown in Fig. 2.6. In contrast, a few hundreds of meters above surface the amplitude of that variation is less than 1°C. Therefore, the diurnal variation of the strength of the inversion must be almost equal to that of the surface temperature itself. When only upper air soundings (generally made once or twice per day at 00 and 12 GMT) are used to determine the inversion strength, it depends partly on the longitude of the station how large a temperature difference $T_u - T_s$ is measured. This has not always been taken into account in studies of the Antarctic inversion. Naturally, such a difficulty does not exist at the SOUTH POLE itself where the diurnal periodic variation of T has to be zero, and in a probably rather small surrounding area where it is insignificant.

The contrast between the average strength of the winter inversion values at VOSTOK and BYRD is remarkable. It should be due, most of all, to the frequent advection of moist, "maritime polar" air masses in the lower half of the troposphere (but above the boundary layer) over Marie Byrd Land, and thus to the appearance of opaque, supercooled waterclouds with relatively large atmospheric radiation $R_L\downarrow$. Similar weather developments are rare over the higher parts of the Antarctic Plateau. Furthermore Byrd Station, situated on terrain

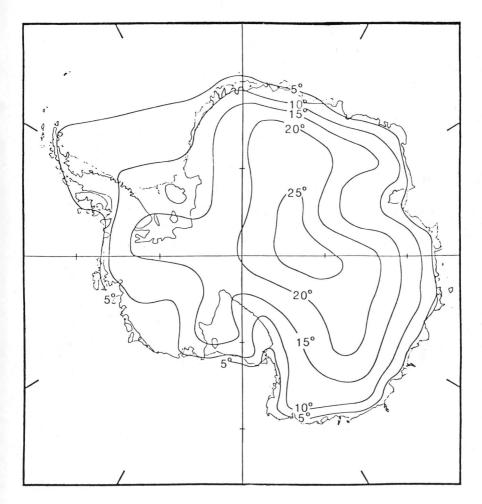

Fig. 2.5. Isolines of the average strength of the surface inversion in the winter. (After Phillpot and Zillman, 1969, with slight modifications.)

steeper sloped than VOSTOK, has generally stronger winds; in the winter half year the mean surface wind speed is 8.7, against 5.3 m/sec at VOSTOK. This phenomenon will be analyzed in detail in the next chapter.

Table 2.11 presents seasonal averages of the strength of the inversion. Fig. 2.6 gives a detailed picture of autumnal conditions at PLATEAU station. Two individual and rather extreme cases appear in Fig. 2.7 and Table 2.12. The former shows the thermal structure of the lower half of the troposphere over VOSTOK, 19 July 1969; mean values for winter and summer have been added only to

32

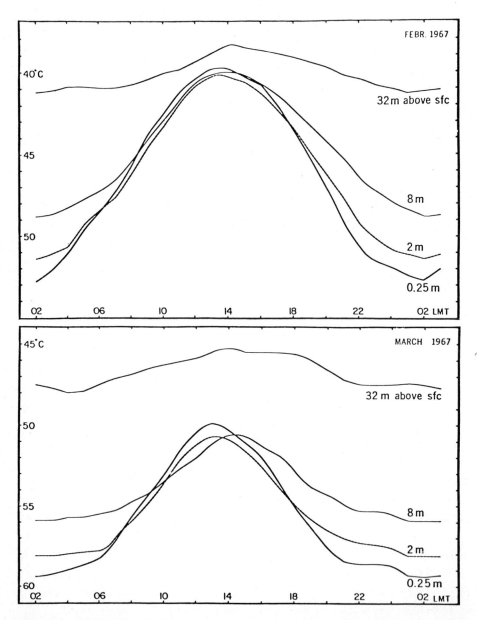

Fig. 2.6. Diurnal variation of temperature at various levels above the surface at PLATEAU Station. The hourly values of each day at each level of a micro-meteorological tower are the average of ten to twenty individual measurements in the first 30 minutes of the hour. Abscissa: local mean time. Upper graph: hourly means for 25 days of February, lower graph for 22 days of March 1967.

TABLE 2.11   Approximate values of the average strength of the surface inversion, $T_u - T_s$; L.T. = local time, Z = Greenwich mean time. T_u stands for the average temperature of the warmest layer anywhere between 300 and 1000 m above surface. For the S-POLE, the temperature at the 650 mb or the 600 mb level was chosen, whichever was higher. In the case of Byrd it was 750 mb, for Hallett 900 mb.

	Summer	Fall	Winter	Spring
	S-POLE,	2835 m	1957-75	
Time of day	Dec., Jan.	Feb., Mar.	Apr. to Sep.	Oct., Nov.
any	1	14	20	11
	VOSTOK, 78.5°S,	3488 m	1958-60	
07 L.T. ≈ 00z	3	15	23	12
	BYRD, 80°S, 1530 m	(summer 12 years, other months 6)		
04 L.T. ≈ 12z	< 1	5	9	6
16 L.T. ≈ 00z	- 3	3	9	3
	HALLETT, 72.3°S, 5 m	1957-64		
	Dec., Jan.	Feb. to Apr.	May to Aug.	Sep. to Nov.
23 L.T. ≈ 12z	- 5	- 4	1	< 1
11 L.T. ≈ 00z	- 7	- 5	1	- 3

facilitate a comparison. It is evident that the conditions for radiative heat loss from the surface must have been good on this as well as the preceding days, with below normal temperatures in the "warmest" layer, weak winds, very small wind shear, and therefore little vertical turbulent mixing in and below that layer. The following day, 20 July, low clouds moved in and the surface temperature rose to -59°C!

The rare event of the complete decay of a winterly inversion between 1 and 32 m above the snow surface at PLATEAU Station is displayed in table form (2.12) because in this case so many measurements at various levels of the micrometeorological tower and free-eye observations are available; all that together would have overloaded a figure. The main point is that here three previously mentioned physical processes must have cooperated to yield an extremely large warming rate, between 21 and 28°C in the twelve hours 18z to 06z; i: advection of warm air above the inversion layer; ii: mechanical vertical mixing produced and maintained by increasing, far above normal, windspeeds, and simultaneously; iii: the longwave radiative heat supply from cloudy air in the "warmest" layer between a few hundred and, say, 3000 m above the surface. Two minor heat sources eventually to be taken into account are the subsurface sensible heat conduction upward, and the latent heating by deposition of H_2O from

34

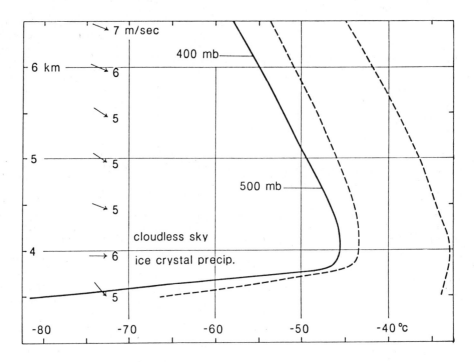

Fig. 2.7. Example of a very strong inversion (36°C in the lowest 400 m) at VOSTOK, 19 July 1969.* At surface, 3488 m elevation, p = 603 mb, T = - 81.7°C, wind 5 m/sec from 315°. The two dashed curves represent the average winter (April to September) and summer (December and January) temperatures, both for the years 1958 to 60 and 1964 to 73.

*Data for July 1969 have been selected from a microfilm copy of USSR antarctic upper air stations' daily records, made available through the World Data Center A, Meteorology, Asheville, N.C., USA.

the air downward, to the cold surface. Both these sources are about one order of magnitude less efficient than the first three (Loewe, 1962; Dalrymple et al., 1966; Weller, 1980).

Table 2.12 leads already into the field of micrometeorology, and Fig. 2.8 does it once more. Both illustrate that in the typical surface inversions of the Antarctic Plateau the strongest increase of temperature with height takes place in the lowest few tens of meters. The more or less realistic mean vertical temperature profile of Fig. 2.8 can well be idealized into an exponential relationship between height and temperature, as formulated in Chapter 3, Section 1.

TABLE 2.12 Destruction of the surface inversion at Plateau Station (79.2°S, 3625 m), 18 - 19 August 1967; the Sun is still below the horizon (in part from Kuhn et al., 1977).

Average values for the 60 minutes ending approx. at		00	06	12	18	00	06 z
Wind speed (m/sec) at	32 m	6.7	8.3	10.4	10.4	9.6	13.8
Wind speed (m/sec) at	8	3.8	4.2	4.5	5.5	6.4	10.9
Wind speed (m/sec) at	1	0.8	1.0	2.0	3.4	4.9	8.7
Difference 32 m - 1 m		5.9	7.3	8.4	7.0	4.7	5.1
Wind direction (°) at	32 m	255	287	289	280	324	283
Wind direction (°) at	8	281	304	308	297	325	276
Wind direction (°) at	1	284	317	333	306	328	279
Change of direction from 1 to 32 m		-29	-30	-44	-26	- 4	4
Temperature (°C) at	32 m	-67.4	-64.5	-63.9	-64.1	-54.8	-43.0
Temperature (°C) at	8	-79.0	-76.0	-70.9	-69.5	-55.2	-42.7
Temperature (°C) at	1	-80.3	-78.2	-75.1	-71.4	-54.9	-42.6
" below snow surface	-0.01	-76.5	-75.8	-73.4	-70.6	-64.9	-62.2
Temp. inversion 32 - 1 m		12.9	13.7	11.2	7.3	0.1	- 0.4
Net radiation (watt/m^2)		- 20	- 19	- 17	- 15	+ 16	+ 18
Cloudiness, tenths, upper		1 Ci	2 Ci	2 Ci	4 Ci	X	X
Cloudiness, tenths, lower		0	2 As	2 As	4 As	10 As	10 As

Additional observations	06 z	hazy moon
	12 z	hazy moon
	18 z	lunar halo
	00 z	drifting snow near surface
	06 z	drift. snow, granular snow precip.

This simplified scheme of the inversion profile, assumed to be valid for undisturbed days of the winter half-year on the antarctic plateau, will probably experience some modification when the acoustic sounder measurements carried out at the SOUTH POLE since 1975 are further developed (Neff and Hall, 1976a and b, 1978; Neff et al., 1977; Hall, 1978; Neff, 1978). Frequently, the echos received have shown a well defined Ground-Based Shear Layer (GBSL) reaching from surface to a height anywhere between 40 and 300 m. When wind and temperature showed little change from day to day, the depth of the GBSL was generally less than 100 m. Such echos indicate that significant turbulence exists, due to a strong vertical wind shear (change of wind and/or direction with height) in a ground-based stable layer. The height of the top of the echo-producing layer tends to be considerably below the layer in which the vertical temperature gradient changes from positive to negative. That appears to be understandable, because the echo measurements indicate qualitatively the occurrences of turbulence; they do not measure temperature gradients. Nevertheless, these two

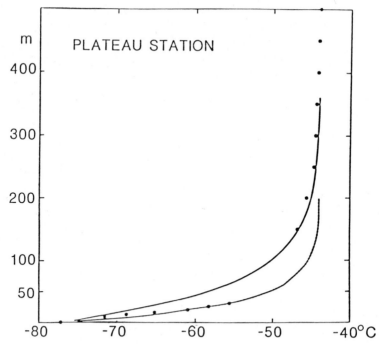

Fig. 2.8. The surface inversion at PLATEAU Station. The heavy dots at various height levels represent the average temperature of ten fair-weather winter days 1967 for which radiosoundings as well as measurements along the 32 m high micrometeorological tower are available. The two exponential curves (computed according to Mahrt and Schwerdtfeger, 1970) approximate the lower (smaller eddy diffusivity K) and the upper part of the temperature profile.

phenomena are not unrelated, and therefore the relatively sharp upper limit of the GBSL seems to be in contradiciton to the smooth curve shown in Fig. 2.8. More complicated echo records, like elevated or multiple echo layers, have been obtained when synoptic or mesoscale fronts, representing changes in density of the air, moved through the South Pole region.

Referring now to the atmospheric layer above a pronounced surface inversion, in Fig. 2.8 approximately above 150 m, many thousands of radiosoundings carried out on the stations in the interior of the continent since the beginning of the IGY in 1957 leave no doubt that normally there is a rather thick layer between, say, 500 and 1500 m (above surface) in which the temperature changes but little with height. Therefore, the often-raised and sometimes unsatisfactorily answered question "how high is the inversion?" is really not a good question at all. Obviously, it is quite irrelevant whether at one level the temperature is half a degree higher than at another neighboring level, or the other way around. Anyway, that could change within a few minutes, but the next sounding will be made 12 hours later, at best. The essential facts are that there exists a layer of considerable thickness where relatively warm and moist air is advected from lower latitudes (Schwerdtfeger, 1968, 1970a,b; Lax and Schwerdtfeger, 1976; Hogan et al., 1982) and also that, by means of longwave radiation, this layer loses heat to the colder air and surface below and the colder air above it. That is confirmed by radiometer soundings carried out at the SOUTH POLE in the sunless part of the year; the heat loss up- and downward happens not only when there is a low cloud deck above the station, but also with cloudless skies as long as there is a strong inversion. Fig. 2.9 illustrates these observed facts.

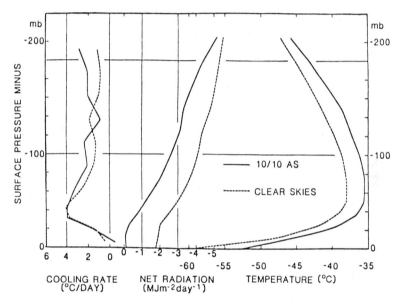

Fig. 2.9. Temperatures, net radiation and cooling rate in the lower troposphere at the SOUTH POLE. Average values for ten cases of overcast and 56 cases of clear sky, winter 1965.

The determination of cooling rates from the difference of the upward and downward directed radiative energy fluxes in the atmosphere above the antarctic Plateau, measured by radiometers attached to the radiosondes, certainly cannot claim a high degree of accuracy. The two radiation fluxes $R_L\uparrow$ and $R_L\downarrow$ can only be rather small at the temperatures of the polar winter, and are not measured simultaneously though in the lowest 200 m they must change rapidly. Nevertheless, an average radiative cooling rate of 4° per day appears to be well acceptable. Bear in mind that the average of the real cooling rate in the respective layer over the plateau in winter is less than 1°C per month. Here are a few data, 500 mb temperatures at SOUTH POLE and VOSTOK, averaged for the same 13 years 1958-60, 64-73.

	April	May	June	July	August	September
S-POLE	-42.3	-43.4	-43.9	-44.8	-45.4	-44.6 °C
VOSTOK	-43.0	-44.2	-44.5	-45.7	-46.0	-46.1

All this means that other than radiative processes, i.e., mainly horizontal advection, adiabatic sinking, and vertical eddy flux of heat, almost entirely compensate for the radiative loss. The above given average 500 mb level temperatures represent the results of complete average energy budgets of an atmospheric layer (not to be mistaken for average radiation budgets of the surface level, which were dealt with earlier). The weather aspect is, of course, quite different. On any individual day, there is no need for the climatic equilibrium to exist; there can be cold air advection, convergence in the boundary layer flow and thus upward vertical motion above it, alternation between a closed cloud deck and a cloudless sky, and so on.

Related to these considerations, there is one more characteristic feature of the strong surface inversion to be mentioned. That is the large variability of the temperature in the equinoctial and more still in the wintermonths from day to day or over even shorter time intervals in the very lowest layer of the inversion (where the "surface" observations are made). In general, the Antarctic Plateau is a slightly sloped terrain with weak undulations, minor wave formations, irregularities due to varying surface roughness, snow drift formations, presence or absence of sastrugi. Such terrain conditions affect the flow of the extremely stable air near the surface. Inevitably, minor divergence and convergence fields can appear. This leads to vertical motions and hence temperature variations that are strongest in the shallow surface boundary layer. A downward motion due to a divergence in the surface wind field, persisting perhaps for only a short while, can easily lead to a temperature rise of a few degrees per hour. Correspondingly, a convergence in the lowest layer brings upward vertical motion and a decrease of the temperature a few tens of meters above surface. In the isothermal layer above the inversion, that is, at a pressure of approximately 600 millibars when referring to the SOUTH POLE region, the thermal effect of vertical motions is much smaller, 1° per 100 meters adiabatically. In a strong inversion it is much more. This is supported by statistical data in Table 2.13, comparing the interdiurnal temperature variations at three levels in July (strong inversion) and January (weak or none).

TABLE 2.13.   24-hourly temperature variations (root-mean-square) ΔT at the SOUTH POLE in January and July, computed for three levels from the daily upper air soundings of 7 years.

	January			July	
Pressure	height above sfc.	ΔT	Pressure	height above sfc.	ΔT
500 mb	2320 m	1.8°C	500 mb	2035 m	2.7°C
600	1030	2.2	600	800	3.2
690	1.5	2.1	676	1.5	6.4

2.5 COLD AIR IN THE SURFACE BOUNDARY LAYER OVER ICE SHELVES

In his first volume on the meteorological results and experiences of the British Antarctic Expedition (Scott's second and last) 1910-1913, G.C. Simpson (1919) describes with some amazement the contrast of temperatures on the Ross Ice Shelf and the adjacent surface of sea ice or land.   Measurements during sledging journeys were usually taken at camps or other longer stops, with hand-held sling thermometers in an aluminum frame, especially developed to serve at very low temperatures. The over-ground height of the measurements must have been about 1.5 m.   In the winter, differences of more than 10°C were found near Cape Armitage over distances of a few miles, and even in summer differences on the order of 5°C were frequent.   Not without compassion one can read about this phenomenon, observed at a winter journey on the ice shelf, also in Cherry Garrard's (1923) famous book "The Worst Journey of the World".

Naturally, the measurements during Scott's expedition could only be made sporadically on the first and last days of sledge travels. When the results of the expedition became known, there were highly respected meteorologists who doubted the data or thought that open sea water must have been close to the rim of the ice shelf. Simpson defended the findings by showing that the temperature differences between the Ross Island stations (Scott's and Shackleton's winter-quarters 1902-04, 1908-09, 1911-12) and Amundsen's station FRAMHEIM on the ice shelf itself (1911-12) 370 km east southeast from Ross Island were of the same magnitude. Meinardus (1938) extended this comparison, making use of the obser-vations of LITTLE AMERICA I and II, located not far from Amundsen's place. Since that time, and even since the IGY, little has been written about the phe-nomenon. Now there is new evidence with the appearance and successful operation of automatic weather stations. Their results agree very well with Simpson's descriptions, and amplify the general picture considerably. Table 2.14 summa-rizes the temperature differences computed from complete series for 24 months of four strategically located stations, i.e., the manned station MCMURDO and the automatic MARBLE POINT on land, and two AWS on the Ross Ice Shelf.   On the last two lines of the table, the differences MCMURDO (27 years) minus LITTLE AMERICA (Meinardus' and 1956 to 1958 data) have been added, as well as MCMURDO minus SCOTT Base (26 years). The latter station is directly exposed to the winds from the ice shelf nearby, while MCMURDO, only 2 km to the west, is somewhat protected by the southern spur of Hut Point Peninsula, and in the second half of summer is closer to open water.

TABLE 2.14. Average differences of monthly mean temperatures of neighboring stations on land and on ice shelf: MCMURDO; AWS MARBLE POINT, 85 km WNW of MCM; AWS FERRELL, 80 km ENE of MCM; AWS MEELY, 110 km SE from MCM and 110 km S from FER. Computed from three-hourly values for the period February 1981 to January 1983, the AWS values measured at about 3 m above surface.

Stations	Types of surface	Summer Nov. to Feb.	Transition Time Mar. and Oct.	Winter Apr. to Sep.
MBP − MCM	Land − Land	0	0.1	0.3°
MCM − FER	Land − Ice	4.7	7.9	10.2°
MCM − MEE	Land − Ice	5.3	7.7	10.2°
MCM − LAM	Land − Ice	4.0	5.0	8.5°
MCM − SCO	Land − Land (SCO closer to ice)	1.8	3.0	3.4°

The agreement between the two first-named stations and again between the two ice shelf stations is quite convincing. The latter show that there is no significant change of temperature along the 110 km south to north line from AWS Meely to AWS Ferrel, on the ice shelf. The average annual surface wind speed at FERRELL is 5.5 m/sec at about 3 m above surface, the prevailing direction from 210°.

The vertical structure of the lowest 1000 m of the atmosphere over the ice shelves in the winter half year can be seen in Fig. 3.14, in Section 3.3.2.

Chapter 3
                                    Dark nights, cold snow,
                                    the winds, they blow,
                                    the snowdrifts grow,
                                    don't let me go.

SURFACE WINDS

In Antarctica the surface winds are decisive for outside working and living conditions. Polar literature and meteorological logbooks abound in testimony to the formidable, rapidly changing forces of nature unleashed in the lowest portion of the atmosphere. The purpose of this chapter is to give insight into the characteristics and causes of the different types of surface winds. Although practical interest might be focussed mostly on the conditions in the more benign part of the year, for the planning of activities which extend into the harsher months it appears appropriate to look at the rapid deterioration of weather in the fall, and the severe conditions during the coldest and darkest time of year.

From the theoretical point of view, the wind conditions near the surface which prevail when the air is coldest and the temperature inversion in the interior strongest are particularly instructive. Due to the topography of the terrain and pronounced stability of the boundary layer air mass, the simple rules of synoptic meteorology regarding the pressure to wind relationship are not adequate to explain the observed winds. Rather, it is the slope of the icy terrain that turns out to be a key factor. Excepting the large ice shelves, there are only singular points on the continent's surface where the slope of the terrain is negligible. Anywhere else on the plateau, the surface winds blow persistently from a direction which depends on the orientation of the ice surface topography. These so-called inversion winds on the gently sloped interior develop in response to the strong radiational cooling of the sloping terrain. Near the base of a steep mountain massif or the coastal escarpments of a plateau, there can be found:

    i. vehement katabatic winds, starting either with a rise of temperature (Foehn-type), or with an arrival of colder air (Borá);

    ii. barrier winds with the air moving essentially parallel to a mountain range or wall;

    iii. frequent calms due to a shelter effect of mountain ranges or promontories in spite of the presence of a sizeable large-scale horizontal pressure gradient;

    iv. in small parts of the coastal regions a funnel effect between mainland and nearby islands appears; it affects wind speed and direction and can extend more than 100 km downwind.

In all such cases the relationship to the synoptic scale pressure field is barely perceptible, if at all. Table 3.1 gives characteristic wind values for

TABLE 3.1 Examples of different types of surface winds: annual mean windspeed
in m/sec, directional constancy q, and relative frequency of calms, C%.
n = number of years used for computation, AWS = automatic weather station;
I.S. = ice shelf.

Type	Description	Station	n	$\overline{V}$ m/sec	q	C%
inversion winds on Plateau	slope > $2 \cdot 10^{-3}$ {	MIZUHO	5	10.3	.96	< 1
		BYRD	14	7.7	.86	1
	slope < $2 \cdot 10^{-3}$ {	S-POLE	16	5.8	.79	2
		VOSTOK	15	5.1	.81	1
	slope near zero	DOME C (AWS)	3	2.9	.45	10*
katabatic	extraordinary	DENISON	2	19.5	.97	0
	ordinary {	MIRNY	17	10.8	.90	2
		MOLO	11	9.9	.85	5
	little	CASEY	10	5.7	.61	13
barrier	on Larsen I.S.	MATIENZO	7	5.0	.73	34
	90 km ESE of MCM	FERRELL (AWS)	2	5.5	.79	6
other winds	near coast on Ice shelf	HALLEY	8	6.5	.62	6
		LIT. AMERICA	6	5.3	.48	5
	on islands	ORCADAS	6	5.0	.28	13
		FARADAY	10	3.6	.21	20

* = percentage of calms for 1980 and 82 only.

a few selected stations with prevailing inversion wind, katabatic wind, or
barrier wind, respectively. In this context the directional constancy of the
wind, called q, is an important parameter. It is defined by the ratio $V_R/\overline{V}$, the
magnitude of the resultant wind (the vector mean of a time series of wind
values at a given place and height) divided by the mean wind speed. A value of
q = 1 would mean that all wind measurements in the respective period of time
indicate the same direction; only the speed could have varied.

Entering the discussion of different types of winds, some cautioning
remarks are in order. According to international agreement, "surface" winds,
reported every three or six hours for the purposes of synoptic meteorology and
equally used for climatological statistics, are supposed to be measured at the
ten meter level above ground. Consequently, either the sensors of a station's

anemometer(s) must really be exposed at that height, or an appropriate extrapolation is to be applied to the measurements. As will be seen (Fig. 3.3), the latter cannot be a precise procedure, though at most stations the differences in height and measured speed will remain small. Unfortunately, there are also some publications of antarctic meteorological records that do not include any respective information.

Where available, the frequency of calms (in % of all observations) has been added to the wind statistics; together with the mean wind speed, that frequency is a good indicator of prevailing working conditions. For instance, the frequency of calms in the winter at station MATIENZO is large; if there is any wind, it tends to be strong, the mean being 8.7 m/sec instead of 5.5 for all observations (Schwerdtfeger and Amaturo, 1979). Also for calms, no great exactness of the statistics can be expected. When wind recorders are used, much depends on the sensitivity of the instrument. Nevertheless, the differences between various groups of stations are large enough to easily surpass the limits of error.

More important obstacles to exact measurements are provided by Nature itself. Mostly in the warmer coastal regions, wet snowfall or even freezing rain can radically affect the form and performance of the wind sensors, and sometimes render immediate repair impossible. Wherever a sufficient amount of snow is available, strong winds can drastically modify the surface, changing it from a smooth plane to a field of sastrugi*, or compiling and moving snow dunes, making the height of the sensor question illusory, and increasing the surface roughness considerably.

---

*

Sastrugi are sharp, unyielding, irregular ridges formed on a snow surface by wind erosion and deposition. Under extreme conditions they can reach heights on the order of a meter. Sastrugi require time and strong winds to form. When the wind turns by, say, 30° and persists in the new direction for some time, two sets of sastrugi will be preserved for some time. Thus a kind of average picture of the stronger and more frequent surface winds and the angular range of directions can be observed. Sastrugi and snow dune orientation, recorded either by traverse parties or aerial photography, have proven to be useful indicators of the prevailing winds in remote regions where no other information exists (Mather, 1960 and 1962, Mather and Miller, 1967).

3.1 INVERSION WINDS

3.1.1 Description

The surface wind field of the Antarctic Plateau is relatively well known. In addition to the records of the meteorological stations, there are measurements with hand-held anemometers and frequent observations of the orientation of the wind-shaped sastrugi. Unavoidably, these additional observations refer mostly to the sunlit part of the years. Nevertheless, all available information, combined with the facts one can read from topographic maps, makes it possible to perceive several wind characteristics which, though not unique, may not be found so pronounced and clear in any other part of the world. Altogether, these winds are a good example of the "Antarctic Atmosphere as a Test Tube for Meteorological Theories" (Lettau, 1971).

The distinctive properties of the surface winds on the plateau, as shown in Fig. 3.1, are:

1) Great directional constancy (as defined on p. 42), decreasing towards the top of the inversion layer (Table 3.2).
2) The vector average of a time series of the surface winds (the "resultant wind") does not point in the direction of the fall line of the terrain; rather, it deviates to the left by 30 to 60°. (For instance, fall-line from S to N, surface wind from SE, as in Table 3.3.)
3) Large wind speed near the surface when the inversion is strong; weak or moderate upper winds have little effect. Inversely, winds increasing with height for lapse conditions[*] (Fig. 3.2).
4) Large speed where the slope of the terrain is pronounced, and the inverse, (Table 3.3).
5) Remarkable changes of the wind direction with height in the lower part of the boundary-layer, when the inversion is strong (Fig. 3.3).

The occurrence of the greatest constancy in the lowest layers (Table 3.2) is of particular interest because it is contrary to what is found in nearly all other climatic zones (Lettau and Schwerdtfeger, 1967). Similar conditions have been found only over the Greenland ice cap (Schwerdtfeger, 1972). Data for PLATEAU Station are not given in Table 3.2 because the number of wind soundings by rawinsonde or pilot balloon carried out at that place is not sufficient to compute a reliable value of the constancy of the upper winds. However, the measurements at various levels along the 32 m high micrometeorological tower erected early in 1967 (Dalrymple and Stroschein, 1977) and still standing there lonely and almost forgotten, make it possible to determine the constancy of the

---

[*] lapse conditions: temperature decreasing with height.

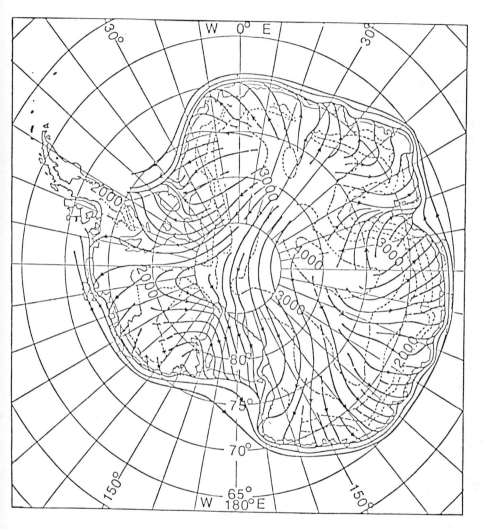

Fig. 3.1. Average pattern of surface winds, inferred from predominant wind frequencies at stations and from traverse records (after Mather and Miller, 1967).

of the winds and its change with height in the lowest 32 meters of the atmosphere, where the temperature inversion is strongest. Pertinent values are shown in Table 3.4.

TABLE 3.2 Directional constancy of the wind near the surface and above the inversion on days with a strong surface inversion; "strong" is arbitrarily defined as >20°C at SOUTH POLE and VOSTOK, >15°C at BYRD station. Data from daily soundings, 1958-68.

Station	South Pole		Vostok	Byrd
Constancy, near surface	0.80	0.80	0.86	0.91
Constancy, at upper level	0.59	0.27	0.55	0.41
Av. sfc. pressure (mb)	680		622	804
Upper level (mb)	650	600	≈ 585[*]	750
Approx. height difference upper level-surface (m)	300	800	500	≈ 500
Average strength of the inversions called "strong" (°C)	22	23	26	19
Number of selected days	426	593	504	323

[*] = av. pressure at the 4,000 m level.

TABLE 3.3 Resultant wind speed (V_r) and direction (D_r), directional constancy (q), and deviation (from the fall line) angle (α), at various places on the plateau; all days.

Station	Data	H	slope	V_r	D_r	q	α	V_r	D_r	q	α
				May to August				Year			
	Years	m	$\cdot 10^{-3}$	m/sec	o		o	m/sec	o		o
PLATEAU ST	3	3625	0.8	4.2	001	.75	49	3.4	335	.67	55
KOMSO	1	3540	1.5	3.0	162	.78	49	3.7	151	.82	59
VOSTOK	15	3488	1.3	4.4	251	.82	34	4.1	243	.81	42
SOUTH POLE	16	2835	1.0[**]	5.3	038	.80	62	4.6	039	.79	61
PIONIERS	1	2740	2.5	9.5	138	.90	48	9.3	131	.92	54
CHARCOT	1	2400	2.5	8.2	164	.87	45	8.6	163	.91	47
MIZUHO	4	2230	3	11.5	099	.97	21	9.9	098	.96	21
BYRD	15	1530	2.5	7.7	013	.88	47	6.6	013	.86	47
SOUTH ICE	1[*]	1350	4	13.5	100	.95	50	10.3	098	.93	52

[*] no data for January

[**] uncertain, see discussion by Mather and Miller (1967, p. 36)

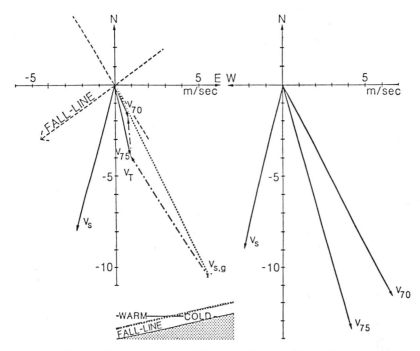

Fig. 3.2 Resultant wind vectors at BYRD station, April to September 1961-1965. At left for days with strong surface inversion, at right for days with lapse conditions. V_S = observed surface wind; $V_{s,g}$ = geostrophic wind at surface level, derived; V_{75} and V_{70} = observed winds at the 750 and 700 mb levels; V_T = thermal wind, dash=dotted.

Some caution is needed when the evidence for the distinctive properties of the inversion winds is examined. No great precision of the numerical values should be expected. This is true in particular for both magnitude and direction of the slope of the terrain surrounding a station on the plateau. Even a slope of $4 \cdot 10^{-3}$, the largest in Table 3.3, cannot be perceived with the naked eye. The values obtained from topographic maps depend, at least at some locations, on the distance over which they are taken and also, of course, on the quality of the maps themselves. As far a the deviation angle α is concerned, the uncertainty of the direction of the resultant surface wind has also to be considered when daily observations only for one or two years are available. Altogether, it might have been as well to give only the nearest multiple of ten for the α values in Table 3.3. In any case, that would have practically no effect on the analytical interpretation of the inversion winds to be given in the following Section. It may still be only a minor exaggeration to say, "the prevailing surface wind direction and speed are so closely related to the direction and steepness of the slope of the terrain, that the former two values can well be estimated if the topography is known, and vice versa".

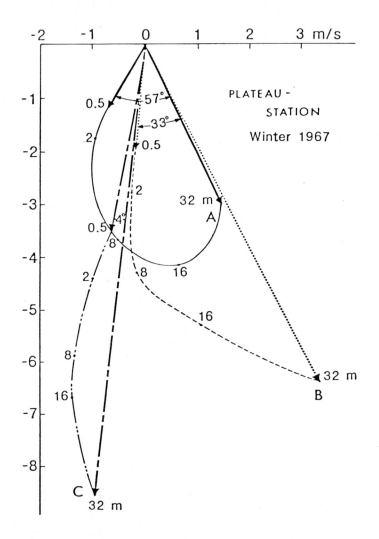

Fig. 3.3 Hodographs of the wind at PLATEAU Station between 0.5 and 32 m, for three different classes of inversion strength:
A: Average of 62 cases of 30 min. mean winds for $(T_{32} - T_{0.5}) > 20°C$;
B: 72 cases with $8° < (T_{32} - T_{0.5}) < 12°C$;
C: 36 cases with $(T_{32} - T_{0.5}) < 3°C$.
Only such cases have been used in which the wind at the 8 m level is from the N – quadrant (the most frequent wind direction at PLATEAU Station).
Time interval between individual 30 min. mean wind profiles: > 6 hours.
(Schwerdtfeger, 1970b).

TABLE 3.4 Mean wind and temperature at 1, 8, and 32 m above the surface at PLATEAU Station, for two periods of different flow characteristics; computed from n 30 minute averages at least six hours apart from another. Period a) describes rather exceptional conditions.

a) Two weeks of highly variable, generally weak winds at 32 m, April 29 - May 12, 1967, n = 43.

H (m)	$\overline{T}$ (°C)	Resultant wind (°) (m/sec)		mean speed (m/sec)	directional constancy
32	-56.9	310°	1.0	5.7	0.17
8	-68.6	013°	1.5	4.1	0.35
1	-73.8	034°	1.0	2.0	0.49

b) One month of persistent, moderate winds at 32 m, July 1 - 31, 1967, n = 80.

H	$\overline{T}$	Resultant wind		mean speed	directional constancy
32	-54.6	356°	6.5	8.3	0.78
8	-63.1	013°	4.8	5.6	0.86
1	-65.0	027°	3.0	3.3	0.91

Note to Table 3.4: Recently published hourly values of temperature and wind measured at several levels of a micrometeorological tower at MIZUHO in 1979 and 1980 show inversions greater than 10° in the lowest 30 m only when the winds are weak, a rare event at that station (Section 3.1.6, Table 3.8 and Fig. 3.8). At wind speeds greater than 10 m/sec, the temperature difference $T_{30} - T_{1m}$ is mostly between 0 and +2°. (Wada et al., 1981, Ohata et al., 1983).

3.1.2 Analysis

The understanding of the prevailing surface winds on the Antarctic Plateau (as well as on the Greenland Inland Ice) at appropriate distance from the coastal escarpments is facilitated by assuming the wind field as a steady state phenomenon. In other words the local changes with time are pretended to be insignificant, so that a vectorial balance of the pressure gradient, Coriolis, and internal friction forces exists for time intervals of several hours to days. Once such an idealization has been used to establish the dominating rules, it is easy to see in what sense some minor modifications, like the diurnal variation of the strength of inversions, affect the resulting winds. Still, it should not be forgotten that the detailed measurements of temperature and wind, carried out at ten levels of the tower at PLATEAU Station, often show large short-time fluctuations in temperature and thus the inherent limitations of the steady state assumption. Miller (1974) evaluated the series of half-hour averages of T for the first six days of August, 1967, as an example of settled winter conditions with clear skies and a very strong inversion (Fig. 3.4). He

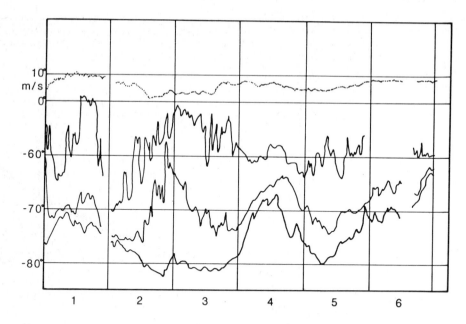

Fig. 3.4. Time series of half hourly averages of temperature and wind speed taken from tower data at PLATEAU Station for the period 1 to 6 August 1967. Solid lines bottom = temperature at 1 meter level, middle = temperature at 12 meter level, top = temperature at 32 meter level, dotted line = wind speed at 20 meter level.

wrote: "It is possible that these fluctuations are symptomatic of a process which contributes to the vertical transports of heat and moisture. Such a process would have length and time scales several orders of magnitude larger than those commonly associated with boundary layer turbulence and could conceivably contribute to a non-gradient component of the turbulent fluxes. Unbroken series of the tower data with time resolutions of the order of minutes are available for periods up to several weeks, and a cross spectral analysis between the series for the different tower levels would give valuable information concerning the frequency spectra of the fluctuations and the vertical variations of their phase and coherence. Such an analysis could better define the role that these phenomena play in the maintenance of the inversion."

Further studies of the profiles of T and V in the lowest tens of meters over PLATEAU Station, and their relation to the eddy transport of momentum, were given by Kuhn et al. (1977), Lettau et al. (1977), and Riordan (1977). The theory of a stable boundary layer over sloped terrain, its dependence on the surface energy budget, and the difficulties to justify the maintenance of a steady state were also treated by Brost and Wyngaard (1978).

Besides the steady state assumption, it is supposed that the inversion layer over the gentle slopes of the interior has a constant vertical extent. All available aerological measurements confirm this assumption within reasonable limits. There is no discontinuous change in the character of the inversion with changing elevation of the underlying surface. As a typical vertical temperature profile (Fig. 2.8, p. 34) strongly suggests, the main temperature increase is in the lower layers. Such a temperature profile can be described by

$$T(z) = T_h - \Delta T \exp(-z/H) \qquad (3.1)$$

where the variable z is the height above surface, T_h the temperature of the free atmosphere above the inversion layer, ΔT the temperature difference between T_h and the temperature near the surface T_s, and H the scale height of the inversion layer, i.e., the height at which the inversion strength is reduced to 1/e of its total value. This kind of approach makes it possible to derive analytical solutions of modified Ekman equations, applicable to conditions in the boundary layer on the Plateau. It was developed by Mahrt (1969, also see Mahrt and Schwerdtfeger, 1970), and modified by Miller (1974) and Lettau (1978).

Another equation needed refers to the relationship between the slope of the terrain, the strength of the inversion, and the resulting "thermal wind", $\vec{V}_T$, of the inversion layer. The latter is defined as the vector difference

$$\vec{V}_T = \vec{V}_{gu} - \vec{V}_{gs} \quad , \qquad (3.2)$$

the two terms on the right-hand side being the geostrophic winds at an upper and a lower level, in this case of few meters above the surface. The 'thermal wind' expresses how the geostrophic wind must change from a lower to a higher level due to the presence of an isobaric temperature gradient in the respective layer. An equation that relates the thermal wind of a sloped inversion layer to the slope of the terrain itself, introduced by Dalrymple et al. (1966), can be written

$$\vec{V}_T = \frac{g}{f} \frac{\Delta T}{\overline{T}} \vec{G} \times \vec{k} \quad , \qquad (3.3)$$

where g is gravity and f the Coriolis parameter, the vector $\vec{G}$ gives the direction and magnitude of the slope line of the terrain, $\overline{T}$ is the mean temperature of the inversion layer, and $\vec{k}$ is the vertical unit vector. The cross product in the equation implies that the thermal wind vector is parallel to the contour lines of the terrain. Only the effect of the sloped inversion layer is here considered; a possibly superimposed isobaric temperature gradient due to the prevailing synoptic situation is neglected. That omission is justified because only the values determined by equation (3.3) for any place on the plateau are always from the same direction, and in the ten winter months rather large, as Table 3.5 shows.

TABLE 3.5   Magnitude of the thermal wind $\vec{V}_T$ of an inversion layer of constant vertical extent and parallel to sloped terrain, as function of the strength of the inversion ΔT and the slope $\vec{G}$.   Assumed: $\bar{T} = 220°K$, $f = 1.4 \cdot 10^{-1}$ sec^{-1}.

$\Delta T(°C)$	$G = 1 \cdot 10^{-3}$	$2 \cdot 10^{-3}$	$3 \cdot 10^{-3}$
10	$V_T = 3.2$	6.4	0.5 m/sec
30	9.5	19.1	28.6 m/sec

Equation (3.2) indicates that the geostrophic winds at an upper level, $\vec{V}_{gu}$, and near the surface, $\vec{V}_{gs}$, cannot be the same unless $\vec{V}_T = 0$, and that $\vec{V}_{gs}$ is opposite in direction and equal in magnitude to $\vec{V}_T$ if $\vec{V}_{gu} = 0$.   Fig. 3.2 gives a realistic example of large (left-hand side) and negligible values (right-hand side) of $\vec{V}_T$, comparing wind vector averages for strong inversion cases and for (rather rare) lapse conditions ($\partial T/\partial z < 0$), at Byrd Station.

For the derivation and application of modified Ekman equations it is assumed that the vector average of a time series of wind data at the upper level (above the boundary layer), $\bar{\vec{V}}_u = \bar{\vec{V}}_{gu}$. In fact, that is a much safer assumption than it would be in individual cases. The same is not true for $\bar{\vec{V}}_s$ and $\bar{\vec{V}}_{gs}$, because there is surface friction (external ground drag and internal Reynolds stress), but equation (3.2) shows the way to determine $\bar{\vec{V}}_{gs}$. The latter is proportional to the mean slope of the isobaric surface at $z = 0$ (the reference height). The magnitude of this slope is on the order of 10^{-4}, under plateau conditions small enough to keep the error small (< 10%) when the horizontal pressure gradient is used instead of the slope of an isobaric surface. In any case, $\bar{\vec{V}}_{gs}$ and another term which takes the baroclinity into account are needed to formulate the modified Ekman equation. In scalar form, u and v are the Cartesian components of $\vec{V}$, and K the coefficient of eddy viscosity:

$$0 = f \ \bar{v}_T \ exp(- z/H) - f \ \bar{v}_g + f\bar{v} + K \frac{\partial^2 \bar{u}}{\partial z^2} \quad \text{and}$$

$$(3.4)$$

$$0 = f \ \bar{u}_T \ exp(- z/H) - f \ \bar{u}_g - f\bar{u} + K \frac{\partial^2 \bar{v}}{\partial z^2} \quad .$$

The analytical solutions to these equations, as formulated by Parish (1980, 1982), equivalent to the original work of Mahrt (1969), are given in appendix A4. Their practical meaning and consequences will be discussed in the following Section.

Prior to that, another approach to the same problem of inversion winds, developed by Parish (1980, 1982), shall be presented. He designed a two-layer model, with the density difference between the sloped bottom level and the equally sloped top level depending on the strength of the inversion. The frictional force is assumed to be proportional to the square of the wind speed, which implies a linear decrease of stress to zero throughout the bottom layer. Such a model can be defined by two scalar equations of motion, to be integrated by computer, with α = specific volume of the air, $\partial p/\partial x$ and $\partial p/\partial y$ the components of the horizontal pressure gradient, and k a friction coefficient equivalent to the ratio c_D/h, drag coefficient over vertical extent of the inversion. The equations are:

$$0 = - g \frac{\Delta T}{T} \frac{\partial z}{\partial x} - \alpha \frac{\partial p}{\partial x} + fv - k \, V \, u \quad \text{and}$$

$$0 = - g \frac{\Delta T}{T} \frac{\partial z}{\partial y} - \alpha \frac{\partial p}{\partial y} - fu - k \, V \, v \quad . \tag{3.5}$$

The solution is shown in Appendix A5.

3.1.3. Comparison of the results of two models

In view of the considerable differences in the design and basic assumptions of the two models, modified Ekman spirals versus a two-layer model, it is satisfactory to find that the main results show much similarity. It is obviously an important question by how much and in what sense the upper level winds $\vec{V}_u$, assumed to equal $\vec{V}_{gu}$, affect the real wind near the surface, $\vec{V}_s$, for various values of the ratio $r_T = V_T/V_{gu}$, with $V_{gu} \neq 0$. Ratio $r_T = 0$ means that the boundary layer is barotropic, the thermal wind zero; $r_T = 2$ represents frequently occurring situations, either with strong inversions over modestly sloped terrain, like VOSTOK, or moderate inversions over steeper slopes, like SOUTH ICE Station; even $r_T = 4$ can easily be found on days with rather weak upper winds. Fig. 3.5 shows that the possible range of directions of the surface wind $\vec{V}_s$ (here computed for 10 m above surface) decreases with increasing ratio r_T. When the same computation of surface winds, as function of the thermal wind of the inversion layer and the (geostrophic) wind above it, is carried through with the two-layer model, the result is quite similar. The decrease of the directional range, equivalent to an increase of the directional constancy, with increasing ratio r_T is in the latter model still more pronounced than in Fig. 3.5 (Parish, 1982). For a given value of r_T the closed curve joining all possible end points of V_s is circular in the case of the exponential temperature profile, elliptic for the other model. Still, there are some inversion wind characteristics for which the differences between the performances of the

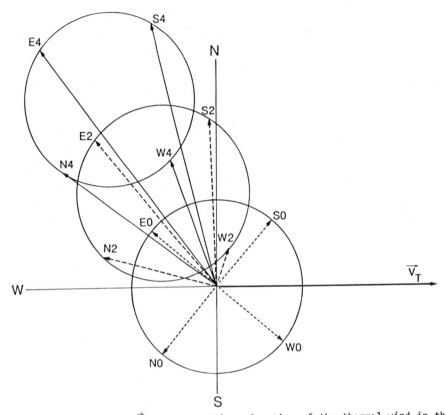

Fig. 3.5. Surface winds $\vec{V}_S$ (10 m level) as function of the thermal wind in the inversion and friction layer $\vec{V}_T$, and of the geostrophic wind above it, $\vec{V}_u$. The speed V_u and the direction of $\vec{V}_T$ are assumed to be constant throughout. Eddy diffusivity $K = 0.1$ m^2/sec, $f = 1.43 \cdot 10^{-4} sec^{-1}$. Each surface wind vector is identified by a letter and a number. The letter indicates the direction from which $\vec{V}_u$ is blowing; the number (0,2,4) represents the ratio V_T/V_u. For any given value of this ratio, the endpoints of the surface wind vectors for all possible directions of $\vec{V}_u$ describe a circle. The graph shows that the greater ratio V_T/V_u, the smaller is the angle including all possible directions of $\vec{V}_S$. There is no scale for the windspeed indicated; only the ratio thermal wind/ upper wind (V_T/V_u) is decisive (Schwerdtfeger, 1970b).

two models appear to be significant. Parish (1980) has given the following summary: i) Given only the strength of the inversion and the slope of the ter- rain, the modified Ekman model provides better surface wind estimates in the interior of the continent than the two-layer model. The temperature profile in these regions often resembles the idealized exponential shape shown in Fig. 2.8. ii) The two-layer model's performance is superior in diagnosing the surface winds in the marginal regions of the antarctic plateau where the slope is gen- erally greater than $2 \cdot 10^{-3}$. The stronger winds prevailing in these regions would tend to mix the boundary layer, producing a temperature profile more reminiscent of the layer model.

An interesting aspect of the model assuming an exponential temperature profile in the inversion layer is presented in Fig. 3.6. It gives the variation of the wind with height in the inversion layer, following Mahrt's (1969) original approach and concentrating on the extreme conditions of $r_T = 4$. For the computation of numerical values, the increase of temperature T with height z in the inversion layer is described by equation (3.1*) instead of (3.1):

$$T(z) = T_h - T \exp(- z \sqrt{f/K}), \qquad\qquad (3.1*)$$

where K is the eddy diffusivity in the inversion-layer $\doteq$ 0.1 m²/sec, and $f = 1.43 \times 10^{-4} sec^{-1}$, adequate for PLATEAU Station (Schwerdtfeger, 1970b). The result is a set of wind spirals -- or hodographs -- which show but little similarity with the original Ekman spiral for a barotropic boundary layer: Complete loops can appear, and it is interesting indeed that the occurrence of such loops in the hodographs of the real boundary layer has been confirmed by Sponholz at PLATEAU Station in 1966 (Sponholz and Schwerdtfeger, 1970) as well as by Kobayashi (1978) at MIZUHO and Adachi (1979) at SYOWA.

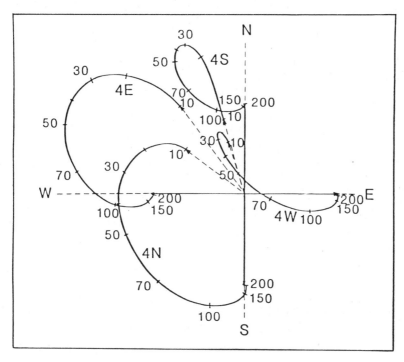

Fig. 3.6. Wind spirals in the boundary-layer computed for winter days with strong surface-inversion and weak winds above it. Assumptions: Eddy diffusivity K = 0.1 m²/sec, V_T/V_u = 4; $\vec{V}_u$ from N,E,S, and W; $\vec{V}_T$ in all four cases from W. Each spiral begins with the wind at the 10 m level; the following numbers indicate the subsequent heights above surface.

3.1.4 Application to the east-antarctic plateau

The models of the inversion winds can be used to produce the mean winter surface wind and streamline pattern over the entire interior of East Antarctica (Parish, 1980). Since the inversion wind concept, as developed above, assumes gentle slopes of the terrain and negligible inertia effects, conditions not fulfilled close to and on the coastal escarpments, the demonstration of the results of the models will not extend downwards beyond the 2000 m (above sea level) contour line. Topographic maps available in 1979 and wintertime mean inversion values as shown in Fig. 2.5 have been applied. In order to examine the effect of the upper wind, V_{gu}, on the surface winds to be computed, the monthly vector-mean geostrophic winds at the 500 mb level (in winter between 4800 and 5000 m, approx.) have been taken from the Southern Hemisphere Upper Air Atlas (van Loon et al., 1971).

Figs. 3.7a and 3.7b (Parish, 1980; 1982) show the results when the possible contribution of the upper wind to the flow in the boundary layer is disregarded. The corresponding maps including the upper wind effect can be found in Parish's (1980) Research Report. It is interesting to note that the the two

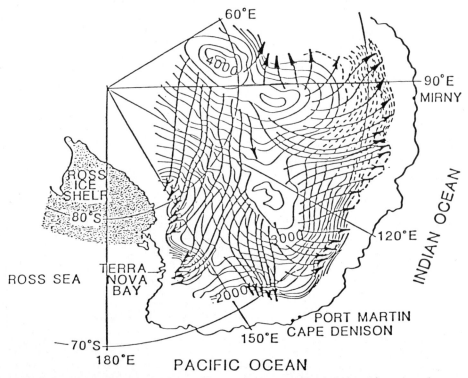

Fig. 3.7a. Time-averaged winter flow pattern over East Antarctica, based on results from the modified Ekman model. The pressure gradient force in the free atmosphere has been neglected.

maps with $V_{gu}=0$, (Figs. 3.7a and 3.7b), give highly satisfactory, "representative estimates of the actual time-averaged winter near-surface airflow over the east antarctic interior", better than those accepting $V_{gu} \neq 0$. Though minor corrections (Schwerdtfeger, 1981) in the upper air maps elaborated in the late 1960s are certainly needed, it is unlikely that such modifications are large enough to alter the above statement. However, it appears possible that the 500 mb reference level of the V_{gu} maps in the Southern Hemisphere Atlas (Taljaard et al., 1969) is just too high, more than 1500 m above the ground near the contourline 3000 m in the maps. Another possibility is that the coupling between the airflow in the extremely stable boundary layer and the flow of the free atmosphere several hundred meters above the top of the inversion is less tight or complete than the two theoretical approaches imply.

This line of reasoning leads to a question regarding the intense, eastward migrating cyclones which almost continuously appear over the Southern Ocean, not far off the coast. The passage of these cyclones can affect the weather along the coast quite profoundly (Tauber, 1960; Streten, 1968; Schwerdtfeger, 1970a), as will be discussed in Chapter 4. However, how far

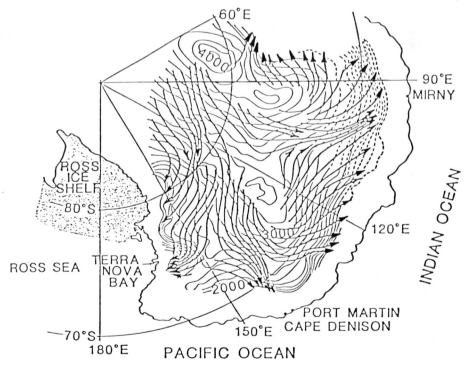

Fig. 3.7b. Time averaged winter flow pattern over East Antarctica, based on results from a two-layer model. As in Fig. 3.7a, the pressure gradient force in the free atmosphere has been neglected.

inland and upslope can the influence of these cyclones be felt? This problem has been taken up by Parish (1980). To estimate the relative contribution of synoptic systems to the surface wind field, he calculated the normalized power spectra of surface pressure for a fair number of coastal and interior stations. The data sets consisted of 1460 six-hourly pressure values for each station for the entire year 1958, only for Cape Denison it was 1913. The relative importance of passing cyclones on the wind regime of a station can then be estimated by determining how large a part of the total variance of the pressure × time series belongs to the "synoptic" time periods between 2.5 and 5.0 days. In the case of the coastal stations, the outstanding feature of the variance spectra is a pronounced maximum in that period, assumed to be due to the transient cyclones. Any such maximum is very weak, or non existent, in the case of the seven stations on the plateau. For all analysed stations, Table 3.6 shows the percentage (of the total variance) attributable to the 2.5 to 5.0 day periods. The entire analysis strongly suggests that the southward penetration of the pressure variations caused by off shore passing cyclones is weak or occurs seldom.

TABLE 3.6. Percentage of the variance of surface pressure responding to the synoptic time period of 2.5 to 5.0 days, for various stations near the coast and in the interior (from Parish, 1980).

5 coastal stations	% of variance	7 interior stations	% of variance
DAVIS	20	VOSTOK	6
OASIS	14	SOVIETSK.	3
MIRNY	17	KOMSOM.	3
CAPE DEN.	13	S-POLE	8
DUMONT D.	15	BYRD	7
		CHARCOT	7
Average	16	PIONERSK.	8
		Average	6

3.1.5 Wind variability and extremes on the plateau

In the first week of December, 1911, Captain Scott's South Pole party camped in the southwestern corner of the Ross Ice Shelf, about 20 km north of the foot of the Beardmore glacier. There they were caught by a most extraordinary storm described in Chapter 4. In the context of the surface winds in the interior of Antarctica, it is of interest that at the time of the storm, Amundsen and his group had already reached the high plateau, at about 3000 m elevation and 500 km south-southeast of Scott's camp. Amundsen (1912) reported: "December 5 gale from the N, the whole plain a mass of blowing snow and thick falling snow; Dec. 6 the same weather, thick snow, sky and plain all one;

Dec. 7 at first like the sixth...., temperatures in these days between -15 and -19°C" (far above normal).

In this case, an intense cyclone had not behaved like a climatologist might have expected, and the surface winds, at least in that part of the plateau, showed no similarity whatever with the nicely explained sloped inversion winds. Obviously, some information has to be given regarding sudden changes and extreme speed of the surface wind. Though by far not as frequent and violent as in other parts of the Antarctic, such changes add considerably to the inclemency of the weather.

Statistics of the highest and the lowest maximum of the windspeed for each month in a period of eleven or more years are shown in Table 3.7. Unfortunately, not all observers, nor the editors of the annual publication "Climatological Data for Antarctic Stations," 1957-75, (U.S. Weather Bureau and successors, 1962-77) seem to have given much attention to the precise definition of the reported values. Terms like "peak gust", "fastest mile" (nautical?) and "daily peak wind" appear indiscriminately, so that some questions remain. In fact, the travelogs of traverses and photographs of imposing fields of sastrugi can tell more than the numbers in Table 3.7. Nevertheless, considering what can happen near the coasts, the relative tranquility in the boundary layer on the higher parts of the plateau becomes evident.

TABLE 3.7. Highest (max) and lowest (min) wind speed maximum (m/sec) recorded for each month at the SOUTH POLE (1957-75, with interruptions), and BYRD Station (Febr. 57 - Jan. 70). Also: Maximum wind speed at VOSTOK (1958-61, 63-73).

SOUTH POLE

	J	F	M	A	M	J	J	A	S	O	N	D	
maxmax	22	20	16	18	24	19	19	23	23	21	17	13	m/sec
minmax	8	10	11	11	12	10	13	13	13	11	10	8	m/sec

BYRD STATION

	J	F	M	A	M	J	J	A	S	O	N	D	
maxmax	21	27	29	32	32	39	37	37	40	31	28	28	m/sec
minmax	12	17	18	19	19	21	26	21	21	23	18	14	m/sec

VOSTOK

	J	F	M	A	M	J	J	A	S	O	N	D	
maxmax	20	16	18	22	15	25	18	16	17	21	16	14	m/sec

The three-year record of PLATEAU Station does not permit a direct comparison with the longer series used for Table 3.7, but Kuhn et al. (1977) give an impressive description of a two-day gale, 13 and 14 November 1967, "in excess of 20 m/sec", maximum >25 m/sec from NNE, blowing snow, visibility zero; "the big storm deposited enormous snow masses in the upwind area of the camp; the surface was raised by approximately 0.5 m".

3.1.6. Additional remarks

During the International Geophysical Year 1957-58, there were eight meteorological research stations installed on the plateau of the continent's interior. Five of them were discontinued at the end of the IGY, a sixth one (BYRD) in 1970. In the meantime, a new station, called PLATEAU Station, was in operation for the three years 1966-68 and brought most valuable data sets, particularly from the extensively instrumented micrometeorological tower (Dalrymple, Kundla, and Stroschein, 1977). At present, 1982, only two stations are continuing their comprehensive observational programs, AMUNDSEN-SCOTT Station at the South Pole, uninterrupted since January 1957, and VOSTOK since 1958 with only the year 1962 inactive. All soundings, near-surface measurements and observations used to develop and test the concept of the inversion wind have been carried out at one or another of these nine stations.

In recent years, a large amount of data have been published in Japan for MIZUHO Station, at 2230 m (JARE REPORTS 2-13, 1974-1983), in operation sporadically since 1971, and continuously since 1976. The slope of the terrain around the station is rather steep, $3 \cdot 10^{-3}$, (Kobayashi and Ishida, 1979); the wind speeds, measured all those years at 4 m and in 1979 and 1980 also at several levels along a 30 m high micrometeorological tower, are correspondingly large. Table 3.8 gives the resultant wind statistics for four years; the directional constancy of the wind is greater than at any other known place. The frequency of days with drifting and blowing snow (Table 3.9) and the reported presence of high sastrugi confirm the strength of the winds (Table 3.8). In the winter, almost 60% of all days have blowing snow and another 30% drifting snow! Furthermore, detailed measurements of the vertical profile in the lowest eight meters (June-December 1972) led to a mean roughness length of 0.24 cm (Sasaki, 1979), due to the presence of high sastrugi at least an order of magnitude larger than the value determined at the SOUTH POLE in 1958 (Dalrymple et al., 1966).

TABLE 3.8. Direction and magnitude of the resultant wind D_R and V_R, average wind speed V, and directional constancy q at MIZUHO STATION, at 4 m above surface, computed from eight 3-hourly observations daily, 1976-79, except February to April only 1977-79.

Month	J	F	M	A	M	J	J	A	S	O	N	D	
D_R	096	100	102	100	101	101	096	097	100	100	092	094°	
V_R	7.1	8.1	9.4	11.0	11.7	11.2	12.1	11.0	11.7	9.2	9.0	7.5	m/sec
V	7.4	8.4	9.8	11.4	12.1	11.6	12.3	11.5	12.2	9.8	9.3	7.7	m/sec
q	.96	.97	.96	.97	.97	.97	.98	.97	.96	.94	.97	.97	
total of days	120	84	93	90	124	120	124	121	120	124	120	124	days

TABLE 3.9. Monthly relative frequency of days with drifting or blowing snow at MIZUHO Station, according to eight 3-hourly observations daily, 1972 and 1976-79, except February to April only 1977-79.
ww = 36 (slight or moderate) and 37 (heavy) drifting snow, vertical extent <2 m; ww = 38 (slight or moderate) and 39 (heavy) blowing snow, vertical extent >2 m. For days with 36 or 37 at some hours and 38 or 39 at others, only 38 - 39 was counted. n = total number of days.

Month	J	F	M	A	M	J	J	A	S	O	N	D	
ww = 36, 37	39	25	27	23	33	23	25	30	35	44	47	45	%
ww = 38, 39	9	18	42	67	60	61	58	53	45	29	24	10	%
n	131	83	93	90	141	150	155	155	150	155	150	155	days.

A comparison of the diurnal variation of the wind speeds[*], and the temperature difference between the top of the towers and 2 m above surface, at MIZUHO and PLATEAU Station is shown in Fig. 3.8. The winds at MIZUHO are strongest when the boundary layer is coldest, between 00 and 06 local time. It appears that these winds are an example of the transition from inversion winds to true katabatic flow. The slope of the terrain increases downhill, so that a complete equilibrium of the acting forces, pressure gradient, frictional, and Coriolis force cannot be attained; the Coriolis force will remain too small as long as the flow of air accelerates where the slope (downward) increases. This interpretation is in agreement with the fact that the deviation angle α (Table 3.3) for MIZUHO is much smaller than for the other stations listed.

There is one more station on the plateau, and this one differs in two ways from the ten other stations where the inversion winds blow. i: It is an automatic weather station (AWS) reporting via satellite, visited only once a year for tests and maintenance. ii: It is the only station on the plateau where there certainly is an inversion, but no inversion wind. Why? There is no, or almost no, slope. The winds at this place (Table 3.10) present a kind of inverse confirmation of the inversion wind theory; they are much weaker and their direction more variable than at any other station on the plateau. The station, called Dome C (74.5°S, 123°E, about 3280 m) is located atop the large ridge extending from the highest elevation of the inland ice (82°S, 80°E, >4000 m) approximately toward northeast; it appears to sit at, or near to, the crest line. Fortunately, the station was in operation for three years, starting 5 February 1980. In January 1983, difficult weather conditions and logistic problems made the annual visitation by C 130 aircraft impossible; apparently disappointed, the automatics ceased to work on the last day of that month.

---

[*] A study of the diurnal variation of the wind at several other stations on ice sheets in summer has been published by Loewe (1974a).

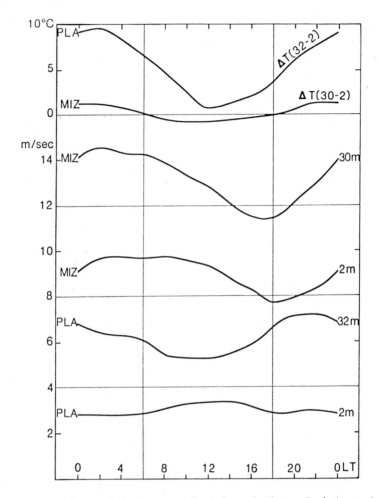

Fig. 3.8. Comparison of the average diurnal variations of wind speeds and vertical temperature differences in the boundary layer at two places in the interior of Antarctica: PLATEAU ST. at 3625 m with a slope of the terrain about $8 \cdot 10^{-4}$, generally weak surface winds, data for 27 days 31 Jan. to 28 Feb. 1967; and MIZUHO ST. at 2230 m, slope about $3 \cdot 10^{-3}$, generally strong surface winds, data of Nov. 1979, 30 days. Availability and completeness of the series of measurements at the two stations determined the selection of the months.

TABLE 3.10. Three years wind statistics for DOME C: mean speed V, vector mean (resultant wind) V_R and D_R, directional constancy q, maximum speed V_X; Units: m/sec. The height of the sensor above surface is 3 m, approx.; n = number of days with three-hourly wind data available. Monthly temperature values are included in Table A3.

	J	F	M	A	M	J	J	A	S	O	N	D	Y
V	2.6	2.8	2.8	3.0	2.5	3.3	3.0	2.7	2.7	3.0	3.5	2.7	2.9
V_R	1.3	1.8	1.3	1.3	0.7	1.4	0.9	1.8	1.3	1.4	2.3	1.5	1.3
D_R	203	204	203	212	188	266	222	182	167	205	187	191	200°
q	.48	.64	.47	.44	.29	.42	.30	.67	.48	.46	.67	.55	.45
n	77	81	92	90	93	90	93	93	90	93	90	92	1074
V_X	9.8	9.4	10.6	10.2	16.0	13.4	15.0	13.9	14.4	14.3	13.2	15.4	16.0

3.2 EAST ANTARCTIC KATABATIC WINDS

3.2.1. Definitions

Gravitational forcing of cold air masses on inclined terrain is an essential requirement both for the formation and maintenance of the inversion winds explained in the first part of this chapter, as well as the katabatic winds which shall now be discussed. In general, the steeper the slope of the terrain, the more the flow of cold, stable air becomes a purely gravity-driven and friction-retarded wind in the boundary layer. It is evident that somewhere between the gentle slopes of the interior and the coastal escarpment there must be a continuous transition, and it appears the MIZUHO station (Tables 3.1, 3.3, and 3.8) is located in such a transition zone. However, a meridional profile of wind and slope, such as that elaborated by Weller (1970), will be more convincing than the wind data of a single station. Measurements carried out at auxiliary stations to the north and south of the permanently manned station MAWSON, and observations of wind and sastrugi orientation farther south, have been combined to show the meridional variation of the surface wind, approximately along 62°E (Table 3.11). Since not all information refers to the same time interval, the wind speed values have been normalized by comparison to the simultaneous data of the permanent station. Altogether, these data as well as a similar investigation in the Mirny region (Tauber, 1960) suggest that on steeper slopes the wind takes on characteristics of purely gravity-driven flow. One has to keep in mind, of course, that the V values are averages of all observations taken during the indicated period, whether or not there were katabatic winds at MAWSON; otherwise the meridional profile of V would be more pronounced.

Notwithstanding, it is desireable to have a dynamic criterion that permits to clearly distinguish katabatic from inversion winds. That is done most easily by making use of the Rossby Number:

$$Ro = \frac{\text{inertial force}}{\text{Coriolis force}} = \frac{U^2/L}{Uf} = \frac{U}{Lf} \qquad\qquad (3.6)$$

where U is a characteristic speed, L a horizontal length typical of the described flow, and f the Coriolis parameter. Rossby numbers Ro < 1 can classify the inversion winds, while Ro > 1 imply that inertial forces are more important, and define the pure katabatic flow.

TABLE 3.11. Mean wind and slope profile north and south of MAWSON
(after Weller, 1969).

No.	Location	Distance from coast, km	Elevation m	V m/sec	Deviation angle from fall line, α	Period of observation
1.	Traverse on sea ice westward, mean	16 N	-	3.5	x	Aug-Oct 1961
2.	Island	11	-	6.6	x	Sep-Nov 61
3.	Island	8.5	-	8.0	x	Jul-Oct 61
4.	On sea ice Micromet. st.	0.5	-	9.5	x	Jun-Nov 65
5.	MAWSON	0	35	9.8	5-10°	1961 and 65
6.	Ice slope micromet.	2 S	~180	10.0	5-10°	Mar 65-Feb66
7.	Ice slope clim. st.	35	~1000	9.9	20	Sep-Dec 61
8.	Southern traverses	110	~1600	7.5	35	Nov-Dec 61
9.	Southern traverses	220	~2400	7.0	55	Nov-Dec 61
10.	Southern traverses	330	~2000	5.6	60	Nov-Dec 61
11.	Southern traverses	440	~2000	3.8	60	Nov-Dec 61
12.	Southern traverses	550	~2000	6.3	60	Nov-Dec 61

It had been shown in the previous section that the idealizing assumption of steady state can be applied, at least for time intervals of several days, to understand the typical inversion winds. That assumption is certainly less admissible in most cases of katabatic winds which are characterized by a high variability of the wind speed even though the directional constancy comes close to 1. This type might be considered as "ordinary" katabatic wind, with high speeds or gusts irregularly alternating with periods of weak winds or even calms, as reported from stations as MAWSON, MIRNY, MOLO, and others.

Probably the most important reason for such a discontinuous flow is the law of supply and demand, combined with the topography of the upwind terrain. A stream of cold air rushing down through a wide glacier valley can continue to do so only if there is a large supply from the respective drainage area where the cold air must have been produced in the lowest layers of the atmosphere. Radiative cooling at the surface of the antarctic plateau is a sure thing for most days between February and November, but this process takes time and varies in intensity, due to changes in cloudiness and strength of the inversion, and thus in the energy budget. Persistent katabatic flow requires a convergence of cold air currents fed by a drainage area of sufficient size and inclination. Over most of the Antarctic interior, the surface winds diverge as the air moves toward the coast. Therefore, persistent katabatic flow can only be the exception (Table 3.1, p. 42).

Besides the supply problem, the intensity of katabatic flow can also be modified by the synoptic pressure pattern and its variations, for instance when the center of a cyclone moves west to east along the east antarctic coast not too far from shore, first hampering, later promoting off-shore air streams. Altogether, the irregular alternating of the speed of katabatic flow really appears to be what one should expect.

Consequently, "extraordinary" katabatic winds are those blowing uninterrupted for days or even weeks, with extremely high speed, practically thwarting the synoptic weather situation. Conditions of this type have been recorded only at two stations in a relatively small sector of the east antarctic coast: CAPE DENISON at about 68°S, 143°E, Mawson's (1915) "Home of the Blizzard" (February 1912 - December 1913); and PORT MARTIN, some 60 km farther W, Loewe's (1972) "Land of Storms" (February 1950 - January 1952).

While the general dynamic aspects of ordinary katabatic wind were explained by A. Defant (1933), Prandtl (1942) and F. Defant (1951), and with reference to the special conditions of the east antarctic coastal regions by Ball (1960), a comprehensive and apparently satisfactory model of the extraordinary winds of CAPE DENISON and PORT MARTIN was brought forth only recently (Parish 1980, 1981).

3.2.2 Ordinary katabatic winds

Representatives of this category are several stations near the coast of East Antarctica at which much in situ research has been carried out (Shaw 1960, Tauber 1960, Rusin 1961, Streten 1963, Burman and Yemshanova 1973). Another station of particular interest, DUMONT D'URVILLE, about 65 km west of the "extraordinary" PORT MARTIN, is located on the Ile des Petrels (archipelago de Pointe Géologie). It is 5 km north of the mainland coast and 1 km west of a northward extending glacier tongue a good part of which broke off in 1977. For a long time it has been conjectured (Mather and Miller 1967) that the surface wind at DUMONT is reduced by the distance from the ice slope and possibly by sheltering due to the surrounding islands, so that at the coast itself wind conditions similar to those of PORT MARTIN might be expected. An automatic weather station, AWS D10, does not confirm such a suggestion. Its complete records for the six winter months of 1982 indicate a mean windspeed $V = 9.2$ m/sec, maximum windspeed 31 m/sec, directional constancy $q = .91$, and frequency of calms $C = 0.5\%$. The station is located only 10 km inland, 270 m above sea

level. In the first weeks after reinstallation, the wind sensor stood 3 m above
the surface, but the snow-cover rises by approximately 1 m per year (pers. comm.
M. Savage).   Fig. 3.9 gives a good example of what can happen even at a place
with "ordinary" katabatic winds. Measurements and specific investigations
carried out at MAWSON and three other stations answer several questions regarding typical features of katabatic winds near the east antarctic coast.

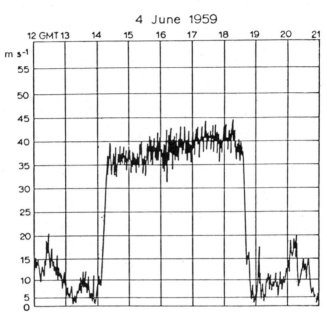

Fig. 3.9.   Example of the sudden onset and cessation of a katabatic storm at
DUMONT D'URVILLE; from Loewe (1972).

i.   The vertical extent of the fast moving air of extreme directional
constancy (q) is, on the average, not far from 300 m, with the wind maximum
itself even lower.   This was found by Streten (1963) who, for the twelve months
March 1960 to February 1961, analyzed 864 pilot balloon soundings which had
reached at least the 900 m level, 275 of them with katabatic wind characteristics, i.e., without much interference from a possibly strong "synoptic" horizontal pressure gradient. To define the latter conditions more sharply, Streten
selected the soundings with the wind speed at the upper level (900 m) less than
5 m/sec and less than the surface wind. His results are confirmed by a slightly
different analysis summarized in Table 3.12. Of the total of 104 selected
cases, 68 have the greatest speed below, 36 at or above the 300 m level.   This
is also in agreement with the studies made by Tauber (1960) at Mirny.

TABLE 3.12. Mean scalar wind speed (m/sec) and resultant wind at MAWSON, at 11, 300 and 900 m above surface, computed for 104 selected soundings March to October 1961. Selection criteria: wind speed at 900 m (above the friction-layer) < 5 m/sec, wind speed at surface > 10 m/sec.

Level	V	V_R	D_R	q
900 m	2.7	1.8	083°	.67
300 m	11.5	11.2	126°	.97
sfc (11 m)	13.2	12.9	136°	.98

ii: Burman and Yemshanova (1973) addressed the same question, but also differentiated between cases of strong katabatic winds presumably not affected by the synoptic scale sea level pressure field, and those cases in which a sizeable pressure gradient tends to contribute an east to west component of the geostrophic wind and therefore modify the wind direction as well as increase the speed in the boundary layer. To this purpose they used the soundings made at MIRNY 1956-69, MOLODEZ 1964-70, and NOVOLAZ 1961-69 at nighttime; the diurnal variation of the wind speed could thus be excluded from consideration. A further necessary selection was to accept as strong katabatic winds only those cases in which the wind direction near the surface was between south and southeast, and the speed greater than 12 m/sec. Fig. 3.10 shows wind speed profiles between surface (10 m) and 2000 m above ground. With the surface wind direction south or south-southeast, taken as indicating little or no contribution by the synoptic situation, the high speed boundary layer at MIRNY and MOLO is very shallow and well defined. When the surface wind blows from southeast, interpreted as due to the presence of a northward directed horizontal pressure gradient, the decrease of wind with height is much less and the speed above the boundary layer is considerably greater. At NOVOLAZ the situation is quite different. Burman and Yemshanova (l.c.) suggest that intense cyclones in appropriate distance offshore are found much more frequently between 0 and 30°E than farther to the east. One might speculate, however, that more conditions than the off shore cyclonic activity must be fulfilled to produce so drastically different vertical wind profiles at NOVOLAZ. For instance, greater roughness of the terrain and stronger vertical entrainment into the downslope rushing air upwind of NOVOLAZ than at MIRNY and MOLODEZ. In many individual cases, particularly at the two latter stations, the term "low level jet", often used in other latitudes, would describe the situation very well. Burman and Yemshanova call it 'mesojet' and apply that name to cases with the surface wind speed >15 m/sec and the (negative) vertical wind shear > 4 m/sec per 100 m. Table 3.13 gives some characteristics for such winds; the differences between neighboring stations are again quite large, and far beyond any reasonable chance limit.

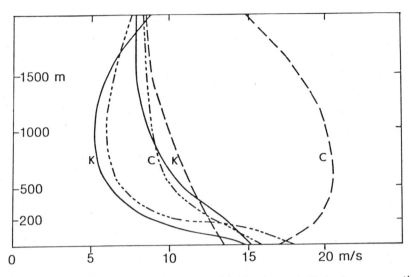

Fig. 3.10. Speed profiles of strong (>12 m/sec at 10 m above ground) katabatic winds, as function of surface wind direction; a strong east-component is interpreted as indicating support by off-shore cyclonic activity. K= pure katabatic; C = cyclonic support.
Solid lines: MIRNY, K = from S and SSE; C = from SE.
Dash-dot lines: MOLO, K = from SSE; C = from SE.
Dashed lines: NOVO, K = from S; C = from SSE or SE.

TABLE 3.13. Frequency of occurrence of katabatic winds (> 12 m/sec) and of the vertical extent of low level jet streams ($\ell.\ell.j.s.$ > 15 m/sec, and negative vertical windshear > $4 \cdot 10^{-2} sec^{-1}$) at three east antarctic stations.

Station	Years of data	Number of cases katab.	$\ell.\ell.j.s.$	<200	200-350	350-750	750-1500	>1500
MIRNY	14	928	496	244	104	40	39	69
ℓ/k			53%					
MOLO	7	487	435	298	113	20	4	--
ℓ/k			89%					
NOVO	9	233	51	9	9	--	11	22
ℓ/k			22%					

iii. The third question refers to the diurnal variation of the wind speed in the sunlit part of the year. At places distant from mountains at lower latitudes, the horizontal pressure gradient and the vertical stability of the boundary layer are decisive for the strength of the surface winds. Normally, the wind maximum occurs in the afternoon hours when the ground is heated and the downward transport of momentum most efficient. At a place at the foot of the escarpment of the antarctic plateau, like MAWSON, the conditions are different. The intensity of the katabatic wind depends on the temperature and thus the density of the downrushing air, the colder -- the denser -- the faster. Hence, the diurnal variation of temperature in the upwind region, at higher elevation, is important for the diurnal variation of the wind speed at the receiving end, i.e., the coastal station. Table 3.14 shows the pronounced diurnal variation of the surface wind speed at MAWSON in January, with the lowest speed in the late afternoon. The periodic diurnal temperature range is only on the order of 4° because the cloudiness is great; the January average amounts to 8/10. More important, however, is what happens in the region from which the cold air comes. At sufficient distance from the Pole, still on the plateau, the radiative diurnal heating leads to much larger diurnal temperature ranges (Fig. 2.6, p. 31). Therefore, the variation of the surface wind speed in summer is not restricted to the coast alone. South of DUMONT D'U., for instance, that was observed in January 1983 as long as a traverse party (planting or restoring automatic weather stations up to 400 km distance from the coast) was on the steepest part of the escarpment where the slope is greater than 10^{-2}, i.e., in the first 40 km approximately. There were very strong winds by night, while late in the morning and more so in the afternoon winds less than 10 m/sec were observed on several days. Higher up, where the slope becomes more gentle, the diurnal variation -- if there was any -- did not come to the observers' attention (personal communication M. Savage). In the winter, of course, the diurnal variation of temperature and wind is negligible.

TABLE 3.14. Diurnal variation of wind speed (m/sec) and temperature (°C) at MAWSON in January and July. 17 year average computed from ANARE Reports, Series D, 1957-68 and data of the Bureau of Met, Melbourne, 1969-73. The time is local time, GMT + 4 hrs.

Month	01	04	07	10	13	16	19	22	L.T.
January	11.1	12.0	11.7	9.9	7.5	6.3	6.3	8.8	m/sec
July	12.9	12.8	12.6	12.5	12.3	12.5	12.8	12.8	m/sec
January	- 1.9	- 2.2	- 1.1	0.7	1.5	1.3	0.6	- 0.8	°C
July	-17.8	-17.6	-17.8	-17.7	-17.6	-17.9	-17.8	-17.8	°C

iv. Another interesting aspect of the east antarctic katabatic winds is the sudden temperature change that often occurs with their onset at a coastal station. This change can be positive or negative, amounting up to several degrees C, but the positive ones are by no means as large as the major Foehn effects in other parts of the Antarctic (Fig. 3.20). Two questions arise: a) how large can the surface temperature difference be between the coastal station where a katabatic flow arrives, and the nearest station on the higher terrain where the flow may initiate? b) once such a stream is on its way downward, what can or must happen to it? There are three pairs of stations which give at least some answer to the first question (Table 3.15).

TABLE 3.15. Temperature differences between coastal and neighboring inland stations, in summer (Dec., Jan.) and winter (Apr. - Sep.).

Station	Elevation	Distance	summer		winter	
			$\Delta T°C$	Lapse rate	$\Delta T°C$	Lapse rate
MIRNY	35 m					
		360 km	21.3	8.0°/km	29.1	10.9°/km
PIONIER	2700 m					
DUD'U.	40 m					
		300 km	21.4	9.1°/km	28.3	12.0°/km
CHARCOT	2400 m					
MOLO	40 m					
*		340 km	17.6	8.0°/km	23.1	10.6°/km
MIZUHO	2230					

*May 1976 through December 1980 for both stations

These data (Table 3.15) indicate that on most days of the winter half-year the difference between the upper and the lower station is greater than 9.7°/km, the value which for the vertical coordinate is known as the dry-adiabatic lapse rate. If the katabatic flow in the winter could be understood simply as a mass transport downward with the potential temperature remaining constant, it would arrive at the lower station colder than the air it replaces. However, the afore-mentioned study by Streten (1963) shows that in the six months May to October 1960 there were only 10 cases with a temperature decrease at onset of the katabatic wind, versus 47 cases with an increase. This discrepancy can be explained by the answer to question b): Above the shallow cold air mass of the boundary layer there lies warmer or, at least, potentially warmer air which can mix with (or be entrained into) the faster moving cold air. The faster the downward flow or the more pronounced the vertical wind shear, the more efficient the process. On the other hand, a quite different process can intervene and complicate the picture: The presence of blowing snow in well-developed katabatic storms. A snow load of 12 gram per m^3, which can easily occur in fast flow, means an increase of the density of the down-rushing air by 1%. Hence, whenever enough loose snow is available along the path of fast flow, there will be a positive feedback. More snow in the air → stronger downward force. Not more than a minor effect could be expected when descending air, sufficiently filled with small snow particles, experiences adiabatic warming only according to the ice-saturation lapse rate, about 9°/km instead of 9.7.

v. One more remarkable feature of the east antarctic katabatic storms is the impressive strength of their maximum wind gusts. The values listed in Table 3.16 speak for themselves, though it must be understood that a high degree of exactness cannot be claimed for any such data. To support this state-ment, the monthly maxima for MAWSON in Table 3.16 have been separated. The first line refers to the 15 years 1957-71, the second only to the two years 1972 and 73. Seven of the twelve monthly maximum gusts occurred in 1973, the highest value of all (84 m/sec in August) in 1972, and only the remaining four in the 15 earlier years. From the statistical point of view, it is most likely that the 17 year series is not homogeneous. One might speculate that in 1971 or 72 a new, more sensitive anemograph has been deployed, or that the old instru-ment has been moved to a place of different exposure.

TABLE 3.16. Maximum wind gusts (m/sec) recorded at four Antarctic stations. Data for MAWSON and CASEY (formerly WILKES) from ANARE Reports, Bureau of Mete-orology, Melbourne; for DUD'U from Monographies de la Météorologie Nationale, 72, Paris 1969; for MOLO from Helbig (1979).

	J	F	M	A	M	J	J	A	S	O	N	D
MAWSON												
1957-71	41	53	52	63	57	61	60	69	62	53	57	53
1972,73	45	47	53	53	77	64	77	84	58	76	76	46
CASEY												
1957-73	52	44	44	56	69	58	58	59	80	62	52	61
DUD'U												
1956-67	57	57	62	68	69	72	62	72	65	73	70	68
MOLO												
1963-74	33	36	46	44	50	51	51	48	48	52	40	40

Finally, one more cautioning remark about deceptive meteorological mea-surements made during stormy weather such as shown in Fig. 3.9: In the scien-tific literature on katabatic flow in the Antarctic, several authors mention pronounced pressure jumps accompanying the onset of strong winds, and try to incorporate this phenomenon into their theory. Unfortunately, no clear distinc-tion has been made between static and dynamic pressure, nor experiments to clarify whether the pressure change has been obtained only inside the station building or in the windswept outdoors too. Therefore, pressure jump effects, though certainly possible from the theoretical point of view, will not be dis-cussed in this context.

3.2.3. Extraordinary katabatic winds

In the meteorological literature of the heroic time of Antarctic exploration one finds many impressive descriptions of the fury of katabatic storms
and the sacrifice of the intrepid observers. Of all these stories, the reports
of Mawson's Australasian Antarctic Expedition 1911-14, and the two winter half-
years spent at Cape Denison in particular, will probably never be surpassed.
The best that now can be done, is to quote verbatim a few paragraphs written by
Mawson himself in his famous "Home of the Blizzard" (1915), and one by C.T.
Madigan (1929). The latter was the expedition's meteorologist who did so much
not only to gather and elaborate the meteorological measurements and observations, but also to slowly convince the many unbelievers.

Mawson wrote:

"Picture drift so dense that daylight comes through dully though, maybe,
the sun shines in a cloudless sky; the drift is hurled, screaming through space
at a hundred miles an hour, and the temperature is below zero Fahrenheit.
Shroud the infuriated elements in the darkness of a polar night, and the blizzard is presented in a severer aspect. A plunge into the writhing storm-whirl
stamps upon the senses an indelible and awful impression seldom equalled in the
whole gamut of natural experience. The world a void, grisly, fierce and appalling. We stumble and struggle through the Stygian gloom; the merciless blast
-- an incubus of vengeance -- stabs, buffets and freezes, the stinging drift
blinds and chokes. In a ruthless grip we realize that we are... poor windle-
straws on the great sullen roaring pool of time...We had found an accursed
country. On the fringe of an unspanned continent along whose gelid coast where
our comrades had made their home -- we knew not where -- we dwelt where the
chill breath of a vast, polar wilderness, quickening to the rushing might of
eternal blizzards, surged to the northern seas."

And Madigan:

"For nine months of the year an almost continuous blizzard rages, and
for weeks on end one can only crawl about outside the shelter of the hut,
unable to see an arm's length owing to the blinding drift snow. After some
practice, the members of the expedition were able to abandon crawling, and
walked on their feet in these torrents of air, 'leaning in the wind'". A
remarkable picture can be found in the British journal 'Weather' (Loewe, 1972);
another one in the Collection of Antarctic Photographs 1910-1916, edited by
J. Boddington (1979).

Sometimes, however, the wind storm abated suddenly for relatively short
time. Again Madigan:

"Considerable interest attaches to these periods of comparative calm and variable winds. Several phenomena were peculiar to them, including small whirl-winds raising snow (like miniature 'willy-willys' with their dust columns in Australia) and also low fracto-cumulus clouds forming rapidly over the coast line, swirling around, drifting north and quickly evaporating. During the calms the wind could frequently be heard roaring on the plateau to the south, and sometimes the snowdrift could be seen whirling down to the coast to the west, showing the coastal calm to be local. Often, too, clouds of drift were observed passing overhead at the 1000 feet level or higher. On several occasions, sledging parties coming in from five miles south reported strong winds at about this level, and walked down into a calm at the hut."

Tables 3.17 and 3.18 give quantitative information about the wind speed observed at Cape Denison. The direction was almost always from south-southeast, $q = 0.97$. The location of the cup anemometer was dictated by the necessity to position the instrument as firm as possible, to stand up against the strongest winds. It was put on a rocky outcrop about 20 m above the level ground near the

TABLE 3.17. Wind speed (m/sec) at Cape Denison, July 1913; values corrected according to recalibration of the Robinson cup anemometer used at the station (Bassett, 1923), published and discussed by Madigan (1929).
DM = daily mean, computed from 24 hourly mean values;
Max = highest hourly mean of indicated day;
Min = lowest hourly mean of indicated day.
All the time, the wind direction was between S and SE.

Day	DM	Max	Min	Day	DM	Max	Min
1	15.9	27	3	17	17.3	22	9
2	32.6	38	24	18	28.3	38	17
3	25.1	33	19	19	24.8	33	15
4	18.1	26	13	20	13.2	21	7
5	31.4	43	17	21	27.9	32	21
6	33.1	42	25	22	20.4	25	16
7	28.9	35	24	23	24.5	34	17
8	29.7	36	24	24	31.3	34	27
9	26.2	32	21	25	31.5	37	22
10	22.9	29	17	26	27.1	34	21
11	30.4	41	10	27	27.2	32	23
12	10.2	26	4	28	15.5	27	5
13	20.1	30	17	29	24.9	30	18
14	19.1	26	13	30	25.5	33	19
15	28.4	30	24	31	30.8	38	25
16	27.4	34	7				
				July 1913 average:	24.8	32.2	16.9

winter quarters (the "hut"), the cups being only about 0.3 m above the recorder box and thus 1.2 m above the rocks. At the return of the expedition to Australia much doubt was expressed, though not by the expedition members, regarding the anemometer's calibration. The instrument was sent to England to be incontestably recalibrated. The result was a reduction of the wind between 5% for the lower speeds and up to 20% for the strongest. All speeds published by Madigan in 1929 are these reduced values. It is interesting to note that a great connoisseur of the polar regions, Fritz Loewe (who had been with Wegener in 1930/31 in Greenland and with the French Expéditions Polaires in 1951 in PORT MARTIN), compared the latter station's data with the CAPE DENISON records. He came to the conclusion the reduction might have been done too drastically, and the DENISON winds of extreme strength were probably even more frequent than reported in the publications of Mawson and Madigan (Loewe, 1972; 1974b).

TABLE 3.18. Wind speed at CAPE DENISON (Madigan, 1929). Number and relative frequency of days on which the 24 hour average wind was below and above various limits.

Limit	Number of days	Percentage
< 4.5 m/sec	1	0.3
< 8.9 m/sec	25	6.9
< 13.4 m/sec	71	19.5
< 17.9 m/sec	130 }365	35.7 }100
> 17.9 m/sec	235	64.3
> 22.4 m/sec	143	39.1
> 26.8 m/sec	63	17.2
> 31.3 m/sec	6	1.6
> 35.8 m/sec	0	0

Total record: 1 Feb. 1912-15 Dec. 1913. Mean of the stormiest month: 24.8 m/sec, July, 1913.
Annual mean wind speed: 19.4 m/sec.
Mean of quietest month: 11.7 m/sec, Feb. 1912
Mean of the stormiest day: 36.0 m/sec, 16 Aug. 1913.
Mean of the stormiest hour: 42.9 m/sec, 6 July 1913.

If there still had been any doubt about such values even after the re-calibration and their detailed publication, it was definitively removed by the experiences and measurements of the afore mentioned French station at PORT MARTIN, about 60 km west-northwest of CAPE DENISON, February 1950 – January 1952 (Barré, 1953; Loewe, 1972). Altogether one might say that the wind conditions at the two places were of approximately equal rigor. The precarious constructions by which the anemometers were kept fast, make it impossible to say at which one it was worse. In 1950, the cups of the PORT MARTIN anemometer were about 7 m above ground. In 1951, the instrument was mounted about 2.5 m on top of the roof of the hut, but soon a big snow drift surrounded the building up to roof's level (Boujon, 1954, Prudhomme and Le Quinio, 1954). Loewe (1972) considers 1951 the year with the more reliable observations: March 21 brought a mean speed of 46.6 m/sec, and the stormiest 24 hours March 21/22 (5 hs to 5 hs) reached 48.5 m/sec. These are probably the strongest daily means obtained near sea level; on a few mountain tops with observations of many years, somewhat higher values have been recorded. July 1913 was the month with the greatest average wind speed at CAPE DENISON, 24.8 m/sec. That value was surpassed at PORT MARTIN in March and April of 1951, with 29.1 and 25.1 m/sec, respectively. As to be expected, the diurnal variation of wind and temperature in the winter months is negligible (Table 3.19). In the short summer, the variations are similar to those at MAWSON (Table 3.14).

The observations made at PORT MARTIN did not only confirm the experiences of Mawson's expedition 40 years earlier; they also helped to eliminate a tempting, though misleading hypothesis about the circumstances causing the extraordinary katabatic winds of CAPE DENISON. Mawson and his men had observed that the strong surface winds inhibited the formation of a solid ice cover on Commonwealth Bay. Kidson (1946), when discussing the meteorological results of the expedition, concluded there should be a causal relationship, in modern terms a "positive feedback", between the temperature contrast - relatively warm sea surface versus cold interior -- and the strength of the offshore winds, understood as a kind of land breeze in a strongly baroclinic boundary layer.

TABLE 3.19. Diurnal variation of wind speed (m/sec) and temperature (°C) in summer (December and January) and winter (May to August) at PORT MARTIN. Data for 1951 (Le Quinio, 1956).

Wind	Local time	00	03	06	09	12	15	18	21 hs
	Summer	14.4	15.7	14.5	12.7	11.3	9.6	8.0	11.2
	Winter	19.6	19.9	19.3	19.8	18.9	18.5	19.2	19.7
Temperature									
	Summer	-3.5	-3.9	-3.3	-1.7	-0.8	-0.4	-0.8	-2.1
	Winter	-18.5	-18.4	-18.4	-18.7	-18.5	-18.6	-18.5	-18.6

It certainly was an attractive idea, only with one serious imperfection: The French expedition found equally strong winds at PORT MARTIN and surroundings as the Australians near CAPE DENISON, but the sea offshore Adélie Land was solidly ice-covered through almost the entire two winters 1950 and 1951, out to an unknown distance.

There is still another place whose wind regime, at least in the six winter months, comes close in strength and persistence of its winds to that of

CAPE DENISON and PORT MARTIN. This spot, only seldom mentioned in the antarctic literature, is part of the east coast of East Antarctica, on 'Terra Nova Bay', 74.8°S and 164.5°E. It is the place where the six members of the Northern Party of Scott's Second Expedition had waited in vain for their ship, the "Terra Nova", and were forced to stay from February to September 1912 under the most exacting conditions (Priestley, 1915). In addition to the precarious food and shelter (in a snow cave) situation, the torment of the party were the brutal westerly winds, reported in detail in Priestley's unpublished diary and only recently analyzed by Bromwich and Kurtz (1982). The authors conclude: "These surface winds collect over a large catchment area on the east antarctic plateau and funnel into "Terra Nova Bay" primarily through the Reeves Glacier. During the winter the katabatic winds blow into the Bay almost continually, thus preventing sea ice from consolidating". The estimated average windspeed of 10 m/sec and the 27% frequency of speeds > Beaufort 7 (> 15 m/sec) outside Priestley's snow cave could suggest a similarity with MAWSON and MIRNY rather than CAPE DENISON. However, it must be taken into account that the six men survived the winter in a shelter they had excavated from a large snow dune. Hence, when they stepped out of their cave, they still came to one of the more protected spots in the entire area.

3.2.4 A numerical model of the Adélie Land winds

Several "favorable" circumstances must be taken into account for the construction of a reasonably realistic model of these winds: i: topography of the upwind terrain, so that a convergent field of motion forms and persists; ii: large size of the drainage area, most of it on the antarctic plateau, and iii: strong radiative cooling in that area, i.e. efficient production of cold boundary layer air. The model should be able to examine if a near balance between supply and demand is possible and an almost continuous (in the, say, eight winter months), vehement flow of cold air can be maintained. Hence, the model must include the equations of motion, mass continuity including the possibility of entrainment, and thermodynamic relationships.

Such a model has been developed by Parish (1980, 1981, 1982), based upon the two-layer scheme described in Section 3.1.2. The assumption of an exponential temperature profile as given by equation (3.1), which proved to be appropriate for the light to moderate winds on the gentle slopes of the plateau, would not lead to a realistic model of the high speed flow of the cold air downward over the rough surface of the escarpment. The details of Parish's model are summarized in Appendix A5. Here it may suffice to show in Fig. 3.11 the result of the model, i.e., the computed flow pattern from the interior toward a height of about 1200 m where the steeper slope of the terrain takes

Fig. 3.11. Winds computed for a model described in Appendix A5, simulating the flow pattern in the hinterland of CAPE DENSION and PORT MARTIN.

over. An area of strong confluence of the cold air drainage currents over the hinterland is clearly evident. Not illustrated here, but certainly important to mention is that changes of the model conditions, like variation of the inversion strength or introduction of a reasonable wind field in the atmosphere above the boundary layer, do not palpably affect the confluence pattern. The size of the drainage area and the supply mechanism via converging currents turn out to be the decisive factors. Most sectors of the coasts of the antarctic continent have smaller drainage areas on the plateau with more di- than con-vergence and, consequently, experience more modest katabatic winds than Adélie Land.

3.2.5 Comment on forecasting

In synoptic and mesoscale analysis one can find developments in the lower layers of the troposphere in which a change of the wind is brought about by a variation of the pressure field, and other weather situations in which the pressure field is modified by a variation of the wind. The latter happens, for instance, when a cold airmass begins to rush down from the Antarctic Plateau, through a wide glacier valley, and finally arrives and advances on an ice shelf, flat coastal ground or coastal fast ice. The pressure field previously existing near sea level will be changed by such an advection of a different airmass, due to the wind. On the other hand, when a heat low (or thermal low) develops on a sunny summer day with calm conditions in the early morning over a desert area, there will be upward vertical motion, outflow in an upper layer, decrease of pressure at surface, then inflow in the lowest layer and finally a cyclonic wind pattern, provided the affected area is large enough and the available time sufficient. In that case, there is first the variation of the pressure field, then the adjustment of the surface winds.

All this can be of interest not only from the theoretical point of view, but also for short range weather forecasting in or near mountainous regions. In the glacier valley → katabatic wind case, information on pressure variations at surface can help but little, or nothing, to give a timely forecast of a new katabatic push of cold air. In the thermal low case, in contrast, perhaps even in appropriate polar regions like the Dry Valleys, a good barograph or hourly barometer readings in the area of concern can be quite useful. The clear distinction of the different character of possible developments like these illuminates the difficulties of forecasting a new outburst of katabatic storms that would mean the end of a quiet period.

3.3 BARRIER WINDS

3.3.1 Remarks on two characteristic regions

A remarkably sharp contrast exists between the wind, temperature, and ice conditions of the two sides of the Antarctic Peninsula. The main reason for that being so, and specifically for the formation of the so-called barrier winds, is the presence of the steep mountain range of the Peninsula, more than 2000 m high south of 68°, between 2000 and 1400 m 68 to 64°S, still nearly 800 m near the tip of the peninsula (63.4°S, 57.0°W). The name barrier winds will be used for cold, low-level winds blowing parallel to extended mountain ranges like the North Slope of the Brooks Range in Alaska (Schwerdtfeger, 1974; Kozo, 1980), and the east side of the Antarctic Peninsula and the Transantarctic Mountains, the latter mostly south of 77°S (Schwerdtfeger, 1975, 1977). The conditions for the formation and persistence of such winds are discussed in

section 3.3.2. Detailed and reliable meteorological evidence for the surprisingly harsh climate of the northwestern corner of the Weddell Sea[*] was one of
the great achievements of Nordenskjöld's Swedish Antarctic Expedition 1901-1903.
There are 20 months of records of the main station SNOW HILL, seven months observations of emergency quarters at Paulet Island, a traverse by ski and dog-
sledge south along the eastside of the Peninsula to 67°S, and many smaller ones
with important geological and biological results (Nordenskjöld 1904, 1911;
Bodman 1910). Up to the last day, when a small Argentine frigate found the men
and brought them back to Buenos Aires, the story is one of the most amazing of
all antarctic expeditions, a master-piece of survival, like a cold fairy tale
with happy ending, everything well documented and recorded, and the meteorological results as valuable as any.

The existence of much milder conditions on the other side of the mountain wall of the peninsula, conditions which in earlier decades had favored the
activities of the whalers and sealers, was confirmed by two French expeditions,
1904 (Rey 1911) and 1909 (Rouch 1914). In later years, a fair number of stations was installed and operated for one or more years, on the South Shetland
Islands and on the west side of the peninsula (for a detailed listing see
Venter (1957), and the annual roster in the Polar Record). East of the climatic divide, however, there were only the British station Hope Bay and the
neighboring Argentine station ESPERANZA, barely 25 km south of the northern tip
of the peninsula (Fig. 3.12).

A decisive change was initiated in 1961 with the installation of the
Argentine station MATIENZO at one of the Seal Nunataks, about 30 km northwest
of Robertson Island, 40 km east of the Nordenskjöld Coast. The Seal Nunataks,
of volcanic origin, protrude through the Larsen Ice Shelf to heights of 200 to
400 m. MATIENZO is located on a spit of land extending eastward from one of the
smaller ones. The observations of this station and their comparison with the
data of the British station FARADAY, at only slightly higher latitude on the
Bellingshausen side, revealed the full magnitude of the effect exerted by the
mountain range on the climate of the area around the peninsula, the wind conditions, and the wind-driven water and ice circulation in the western Weddell Sea
(Schwerdtfeger 1975 and 1979). Fig. 3.13 shows the mean monthly temperatures
at various places on both sides of the peninsula. The annual wind speed, directional constancy and relative frequency of calms at the two stations MATIENZO
and FARADAY are included in Table 3.1; more information on MATIENZO winds in
Table 3.20.

---

[*]The name Weddell Sea is here understood to apply to the water body south of a
line from Joinville Island (63°S, 53°W) to Cape Norvegia (71°S, 12°W).

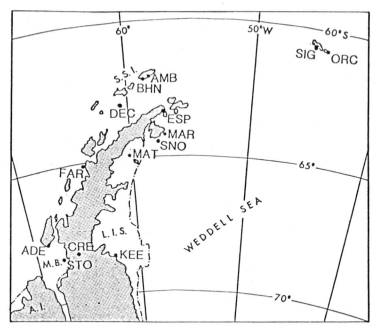

Fig. 3.12. Location of some meteorological stations in the region of the Antarcic Peninsula. S.S.I. = South Shetland Islands; L.I.S. = Larsen Ice Shelf; M.B. = Marguerite Bay; A.I. = Alexander Island. The waters between S.S.I. and the northernmost part of the Antarctic Peninsula are called Bransfield Strait. Stations CRE (68.1°S, 66.4°W, 1768 m) and KEE (68.9°S, 63.2°W, 23 m) existed only in the spring of 1947. The other stations are listed in Table A1.

On the side of the Bellingshausen Sea, stations at latitudes higher than 65°S had been in operation intermittently since 1936 [British Graham Land Expedition (Stephensen, 1938)] on islands close to the shore of the peninsula, in Marguerite Bay. Of particular interest are the U.S. Antarctic Service Expedition 1940-41 and the Ronne Antarctic Research Expedition 1947-48; the former because it included, in addition to the main station on Stonington Island, the first-ever mountain station in the Antarctic (Dorsey, 1945), located on the crest of the peninsula at 1637 m above sea level east of Stonington Island November and December 1940. The Ronne Expedition, in turn, maintained stations not only on Stonington Island (1 year) and on the crest at 1768 m (September-November 1947), but also a third station on the other side of the mountains at Cape Keeler (68.9°S, 63.2°W, 23 m), in October and November 1947. Up to the time of this writing, these two months are the only time in which synoptic observations have been made at fixed places on both sides of the peninsula, south of 65°S. That series of observations, as short as it is, proves to be valuable for the understanding, and hence the forecasting, of the wind conditions along the east side of the peninsula. On the west side, there

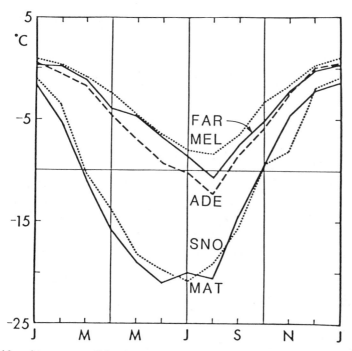

Fig. 3.13. Average monthly temperatures on the west side (upper three) and on the east side (lower two) of the Antarctic Peninsula.

have been intermittently, since 1947, several meteorological stations south of 67°S: FOSSIL BLUFF in George VI Sound; SAN MARTIN and STONINGTON in Marguerite Bay; HORSESHOE ISLAND, ADELAIDE ISLAND and ROTHERA. Their location is given in Table A1, more information by Schwerdtfeger et al. (1959) and Schwerdtfeger and Amaturo (1979).

The second region whose topography strongly favors the formation and frequent occurrence of barrier winds is the western part of the Ross Ice Shelf. To the west of that region, the imposing Transantarctic Mountains and behind them the high plateau make any westward advance of stable air in the lowest 2000 m (at least) of the atmosphere simply impossible.

It is a strange fact that in the scientific meteorological literature of the past thirty years there have appeared on the order of hundred papers about the flow of air over mountains, but probably not more than five about what happens to cold and stable air that moves toward a mountain barrier and accumulates on the windward side. Questions of this kind had occupied the attention of our professional ancestors 70 to 80 years back. The piling up of cold air masses approaching the central range of the Alps from the north and the consequent surface pressure increase have been analysed in all detail by Ficker

(1906). A shorter description and two sketches of a mountain range with the sloped cold air mass on the windward side and the pertinent pressure field can also be found in Alfred Wegener's (1911, Figs. 60 and 61) textbook "Thermodynamik der Atmosphäre". Such a development was probably also on the mind of G.C. Simpson (1919) when he elaborated the meteorological observations made at Scott's winter quarters (CAPE EVANS) and at other places on Ross Island. Furthermore, Simpson made use of the several minor traverses whose main purpose was to deposit food and fuel at various well-marked places to be used in the coming summer by the Pole-party, and other traverses with scientific goals. All this was the only meteorological information for that area between 1912 and 1956. From then on, the regular activities of MCMURDO and SCOTT stations, the latter only two kilometers to the east of the former, brought uninterrupted series of observations. For the summer 1976/77, a fair number of glaciologists active at the various work points of the Ross Ice Shelf Project stepped in as meteorological observers at their respective places. A few valuable series (Schwerdtfeger, 1979) were produced though, unavoidably, only for the less characteristic summer months. The real salvation came in 1980 with the installation of the first workable U.S. Automatic Weather Stations (AWS), see pages 8 and 9, Fig. 1.4. After the first three years of these new tools there is not the least doubt that the measurements of the AWS deployed south, southeast, and east of MCMURDO (Stearns and Savage, 1981; Stearns et al., 1982) confirm very well the flow pattern derived by Simpson (1919) from sporadic observations and discussed by Schwerdtfeger (1977) for the northwestern part of the Ice Shelf.

3.3.2 The origin of barrier winds

It does not need explanation that strong winds must blow northward along the east coast of the Antarctic Peninsula whenever an intense low pressure system is situated over the central Weddell Sea, and correspondingly along the east flank of the Transantarctic Mountains, when a deep cyclone or trough lies over the Ross Ice Shelf. The only serious problem is that adequately intense pressure systems of that kind seldom occur, while the strong southerly winds along the mountainous coasts are by far the most frequent winds in the two respective regions. Consequently, there must be another type of (sea level) pressure field and flow pattern which forces the air in the lower 1000 meters of the atmosphere near the mountain barriers into a range-parallel motion.

The two regions briefly described in the preceding section lie far south of the circumpolar low pressure trough. Correspondingly, easterly winds prevail in the lower layers of the troposphere of the higher latitudes as long as no major obstacles are in the way. The air in the lowest 500 to 1000 m is very stable. The potential temperature at 1000 m in the two summer months December and January is between 5 and 10°C higher than at the surface, and in the rest of the year 10 to 20°C higher. It is important to note that the lowest tens of meters of air over the ice shelves tend to be considerably colder than the air over neighboring land. This is discussed in Section 2.5.

Fig. 3.14 compares the results of winter half year soundings in the lowest 1000 m at two stations located on an ice shelf, one set of soundings carried out aboard ship drifting with thick sea ice, and one station (MCM) on land only a few kilometers off the ice shelf. Point AWS shows the mean winter temperature 1980 and 81 at two automatic weather stations, about 100 km southeast of Ross Island. The precarious location of ELLSWORTH Station on a protrusion of the Filchner Ice Shelf (pp. 4 and 39) should be taken into account for an interpretation of the profile ELL.

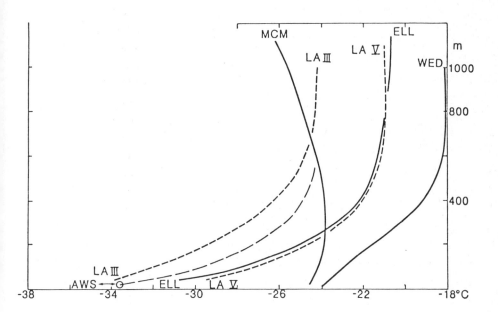

Fig. 3.14. Winter average profiles of the inversion layer over ice surfaces; comparison with MCMURDO. April to September = winter.
L.A. III: LITTLE AMERICA III, 78.5°S, 115 radiosoundings 1940 (Court, 1949).
L.A. V : LITTLE AMERICA V, 78.2°S, daily soundings 1956-1958.
ELL   : ELLSWORTH, 77.7°S, daily soundings 1957, 1958, 1962.
WED   : Ice-covered Weddell Sea between 72°S/41°W and 69°S/45°W, 50 kite-
          soundings May-July 1912 (drift of the DEUTSCHLAND), Barkow (1924).
MCM   : MCMURDO, 77.9°S, on land about 5 km north of the Ross Ice Shelf;
          daily soundings 1965-1969.
A W S   : refers to the little circle at -33.6°C, indicating the average
          temperature of the automatic stations 8907 (100 km ESE from MCM)
          and 8915 (105 km SE) measured about 3 m above surface, winter 1980
          and 81.

The dashed line outgoing from the AWS circle is only an estimate of the average inversion over the northwestern part of the Ross Ice Shelf.

After this digression to the temperature inversion over an ice surface, attention will be directed to the motion of this air. East to west wind components prevail over the eastern and central parts of the two large embayments of the continent, i: Filchner and Ronne Ice Shelves with Weddell Sea south of, say, 65°S, and ii: Ross Ice Shelf, in winter with the southern Ross Sea. This motion, however, cannot continue westward indefinitely. As the very stable air approaches the steep, south to north extended mountain ranges, it has little chance, or none at all, to be forced over the barrier. What will happen is a damming up against the west side of the mountains, thereby producing an east to west upward sloping upper surface of the inversion layer, and increasing the surface pressure along the east side of the mountains.

In the case of the Antarctic Peninsula, some flow over the crest is possible when the piling up reaches high enough (Section 3.3.5). In the Transantarctic Mountains, backed up by the ice masses of the plateau, there is, of course, no overflow even under the most favorable conditions. On the contrary, in some glacier valleys and otherwise appropriate terrain configuration, katabatic flow of potentially colder air (occurring mostly in the winter) can undercut the upslope rising air of the stable boundary layer. Such a complication will be discussed in Section 3.4.4.

The inversion layer over ice shelves and ice-covered waters will become inclined when it is moving toward a widely extending mountain range, with a slope line approximately rectangular to the long axis of the obstacle. This is, of course, not a new discovery. A similar stratification and piling up of stable cold air masses, sometimes recognizable by a shallow stratocumulus cloud deck above the inversion and moving from the north toward the central range of the Alps, has been thoroughly analyzed by v. Ficker (1906). The same phenomenon can be observed on the east side of the southern Andes from a minor top of the Precordillera, and at many other places. The known elevation of passes and peaks serves as height scale.

Under such conditions, there must be an isobaric temperature gradient (lower temperatures near the mountains) a few hundred meters above sea level, and hence a thermal wind parallel to the mountain range. Fig. 3.15 shows the principle involved. If the temperature contrast between the cold air near the surface and the warm air above it is denoted by ΔT and the slope of the upper surface of the cold air layer by $\vec{G}$, the resulting thermal wind $\vec{V}_T$ can be estimated from equation (3.3) on page 51. The magnitude of the upslope vector $\vec{G}$ could be measured by airplane, though apparently no such flights have yet been carried out in polar regions. For cases of cold air moving in the undisturbed up-wind region at a right angle to a steep mountain wall farther downwind, it may be on the order of 1/100. This estimate is based upon the assumption that

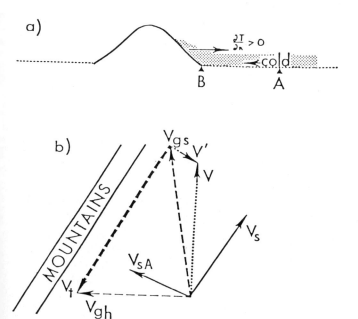

Fig. 3.15. Schematic illustration of the flow of very stable air toward and along a mountain range (Southern Hemisphere). a) Stratification and isobaric temperature ascendant $\partial T/\partial n$. b) Wind vector diagram: V_{sA} represents the surface wind over ice at point A in a), not affected by the mountains. All other wind vectors refer to the coastal region, near point B. V_{gh} = geostrophic wind above the cold-air boundary layer; V_t = thermal wind of this layer; V_{gs} = geostrophic wind at the surface level; V = wind at surface level without friction; V_s = surface wind; V' = ageostrophic wind component due to curvature of trajectories; this component can be larger than indicated in the graph.

the mountain-parallel component of the geostrophic wind (equal in magnitude and opposite in direction to the thermal wind) is on the order of 20 m/s, an appropriate value for strong barrier winds. Since the geostrophic wind at a lower level equals the geostrophic wind at an upper level minus the thermal wind (equation 3.2), the surface wind $\vec{V}_s$ must turn to the right (southern hemisphere) as the cold air approaches the mountains. If the upper wind is weak in comparison to a strong thermal wind, the surface wind must also be strong and tend to blow parallel to the range.

Another way to understand the conditions on the upwind side of a mountain range refers to the atmospheric pressure field at various levels (Malberg, 1967). Both approaches, via thermal wind or pressure gradients, assume that a steady state can establish itself, with an equilibrium between pressure gradient-, Coriolis, frictional, and inertial accelerations. This is in agreement of the fact that in many cases strong, mountain-parallel surface winds persist for several days. The justification of the entire line of reasoning as sketched in Fig. 3.15 has recently been strengthened by Parish (1983) by means of numerical experiments using a two-dimensional primitive equation model for a cross section of the terrain of the Antarctic Peninsula.

3.3.3 The winds on the east side of the Antarctic Peninsula

For a meaningful analysis of the surface wind conditions in the western Weddell Sea region, here represented by the station MATIENZO, the generally applied type of statistics, determining V, V_R, D_R and q, has a serious shortcoming. The reason is that in one and the same month there may be southwest winds which temporarily can turn into west winds with temperatures of -20 to -30°, as well as west winds varying between WSW to NW with temperatures near the freezing point. In the first case, one deals either with barrier winds, or winds due to the presence of a low pressure system, the former carrying continuously, the latter intermittently cold boundary layer air from higher latitudes toward the station. In the second case, there are foehn winds which with sinking motion bring air from higher elevations down to the station level. Examination of the three-hourly wind and temperature data for a twelve months period has indicated that a clear distinction between these two different classes of wind can easily be made in most cases, provided only winds with speed > 5 m/sec are taken into account. In the few instances of borderline conditions, the change of temperature accompanying the increase of wind speed can be used as criterion: temperature decrease characterizes barrier winds, and an increase characterizes foehn winds.

The restriction to cases with wind speeds greater than 5 m/sec not only helps to separate foehn-affected winds from barrier winds, it also makes the frequency values of the different wind classes more applicable to considerations of wind stress and inherent effects on ice; the weaker winds have less directional constancy, and exert only negligible stress.

As Table 3.20 indicates for a period of 2920 three-hourly observations (12 months), there are only two significant classes of winds on the Larsen Ice Shelf east of the Antarctic Peninsula. i: cold barrier winds including a few cases with winds due to an intense cyclone situated in the central part of the Weddell Sea; and ii: foehn winds attributable to strong westerly or northwesterly winds when the circumpolar low pressure trough west of the Peninsula lies south of, say, 70°S. These winds are described in Section 3.4.1.

When one divides the entire twelve months period into parts of unequal lengths and includes results obtained for other years because the "seasonal" variations can thus be shown much more clearly, the following picture develops (Schwerdtfeger and Amaturo, 1979). In the western Weddell Sea region, already

TABLE 3.20.   Surface winds at MATIENZO.
   a) Relative frequency of four classes of winds, for three   minimum
      speed limits.   March 1968 - February 1969; total number of all
      three-hourly observations: 2920.
   b) Maximum gusts during these twelve months

a) class of winds

class of winds		speed >5	>10	>15 m/sec
S-W, cold barrier including cyclonic winds		78	79	77%
W-NNW, warm, foehn-affected		14	20	22%
Other winds	N - ENE	6.5	1	1%
	E - SSE	1.5	0.1	0%
Number of observations		1461	827	488

b) Maximum gusts, barrier winds: SW, 86 m/sec, -17°C,   7 Aug. 1968
                    foehn winds: WNW, 41 m/sec, +4°   27 Sep. 1968

Note: At the first glance, the maximum gust value of 86 m/sec seems to be
unlikely strong. It might therefore be added that the five minute average
wind speed recorded two hours prior to the fierce gust amounted to 39, one hour
after it to 46 m/sec. Through all of the winter 1968 the observers have made
careful annotations of special events. It is not known what type of anemograph
has been used.

the month of February brings in most years several events of strong cold air
advection from the south (Bodman, 1910; Schwerdtfeger and Komro, 1978). The
twelve years temperature series of MATIENZO (1962-72 and 75) shows that the average decrease from January to February amounts to 4.0°C; in the same years,
200 km to the west, at FARADAY, it is only 0.1°.   It appears that the seven
months period February to August has the highest frequency of cold barrier
winds, about 40% of all observations, versus 3% for foehn winds.   In contrast,
for the two months September and October these frequencies are 32% and 26%.
Such a change from winter to spring fits into general climatological experience.
In those two spring months the anticyclonic pressure cell over the southwestern
South Atlantic tends to lie farther south than in the rest of the year, so that
the low pressure trough line between mid-latitude westerlies and polar easterlies in the peninsula region can frequently be found south of 65°S. The alternation of the two important wind classes, cold from south, and warm from west,
as frequently observed in spring, is also of interest with respect to the
transport of ice toward lower latitudes. The foehn winds force the ice away
from the edge of the Larsen Ice Shelf and the islands south of the northeastern
end of the Antarctic Peninsula (Kyle and Schwerdtfeger, 1974, Colvill, 1977);
the barrier winds then drive the ice northeastward (Gordon and Taylor, 1975;
Schwerdtfeger, 1979).
     Since up to here the explanation of the barrier winds in the north-
western part of the Weddell Sea is based mainly upon the records of the station

MATIENZO, some supporting evidence shall be added in the form of comparable observations for the cold season at Nordenskjöld's station SNOW HILL (160 km northeast of MATIENZO, 20 km southwest of the present station MARAMBIO) in 1902 and 1903. For the eight months March to October hourly values of wind and temperature are available for both years (Bodman, 1910). There are 88 days in which the 24 hour average of wind speed (measured at 2 m above surface, 4 m above sea level) was greater than 15 m/sec and the direction from southwest. Table 3.21 gives the mean speed and mean temperature of these days. It shows that the air masses moving north and northeast along the eastern slopes of the Peninsula have temperatures which correspond to much higher latitudes, that is, to the southern regions of the Weddell Sea. The two longest periods of consecutive days with barrier winds > 15 m/sec are also listed. Both cases belong to the year 1902 which must have been a year of particularly severe weather conditions, with 54 of the 88 days defined above.

TABLE 3.21. Strong, cold barrier winds at SNOW HILL in 1902 and 1903, March to October. Average wind speed and temperature for 88 days, each of them with a 24-hour average wind speed > 15 m/sec.

Month	M	A	M	J	J	A	S	O	
Number of such days	17	10	7	11	13	11	10	9	
Mean wind speed	19	18	21	18	20	21	19	18	m/sec
Mean temperature	-13.8	-18.6	-22.7	-24.3	-24.3	-24.6	-21.0	-18.1	°C

Longest periods of consecutive days with cold SW winds > 15 m/sec:

1 - 7 June 1902: V = 20 m/sec, T = - 25.6, T_{max} = - 23.8, T_{min} = - 27.1 °C;
15 - 21 July 1902:   21 m/sec,   - 25.6   - 23.0   - 28.1

It is obvious that the stress exerted by such surface winds is the main force which drives ice and cold water masses from the northwestern Weddell Sea into the west to east circumpolar current of the southern ocean. In recent years, this can directly be seen in the weekly ice maps of the U.S. Navy's Fleet Weather Facility (1971-present). The climatic consequences of this ice transport will be discussed in Chapter 6.

The records of SNOW HILL and MATIENZO indicate that periods of several days with uninterrupted strong south-southwest winds and low temperatures are quite frequent. That gives some additional information regarding the type of the prevailing synoptic situations. If the winds were maintained by a stationary cyclone over the central Weddell Sea, warmer and moister air would be carried around the low pressure center, at least intermittently. In contrast, moderate easterly winds over the central Weddell Sea can persist easily for a few days without admitting the influx of warmer air.

Before turning to an individual barrier wind event in the next section, two characteristic features of the winds along the east side of the peninsula, based on six year statistics for MATIENZO, must still be mentioned:

a) Very strong wind gusts have been measured, in particular in the months March through October; the maximum in a barrier wind situation 86 m/sec, in a foehn storm 61 m/sec (in June 1970).

b) The frequency of calms is extremely high; the annual mean is 34%; only about 15% of all observations remain for weak winds (< 5 m/sec). It is likely, though not yet proved by observations, that the frequency of calms decreases southward.

3.3.4 An extended period of barrier winds

The first eight days of May 1968 present a good example of barrier winds. This time interval has been chosen mainly because there can be not much doubt, in spite of the lack of data from the Weddell Sea proper, that a broad stream from the east carried cold, stable air masses across the mostly ice-covered sea toward the Antarctic Peninsula. That is strongly suggested, to say the least, by the fact that the eight-day average sea level pressure difference near 40°W, BELGRANO at 77.8°S minus ORCADAS at 60.7°S, amounts to 25 mb, the individual values varying between 15 and 35 mb. A 25 mb difference over the distance of 1900 km corresponds to a value of 7.5 m/sec (at density 1.3 kg/m^3) for the com-ponent of the geostrophic wind from 085°. The winds at MATIENZO are not excep-tionally strong in that period; only on two of the eight days does the speed rise to the relatively modest value of 22 m/sec. However, the directional con-stancy of the winds is almost total (see Table 3.22), and that is so not only at the key station MATIENZO, but also at the other stations to the northeast and, more important, at station STO at 68°S on the other side of the mountains where the wind blows continuously from the ESE.

The high persistence of the wind direction makes it meteorologically meaningful to draw an "average synoptic" map for the entire eight-day period, shown in Fig. 3.16. Though the pressure field farther to the west of the penin-sula remains unknown, an unavoidable conclusion is that a narrow high pressure ridge exists east of the crestline of the mountain range. That is exactly what must be found when cold stable air, moving westward from the central Weddell Sea, is piling up on the eastern flanks. Such a situation also favors the oc-currence of katabatic winds on the lee side of the mountains. It must be noted, however, that it cannot be the air originally near the ice cover on the Weddell Sea, with a temperature of -19°C or lower. The air that arrives at Marguerite Bay with about -8° must also come from the Weddell Sea side, but from a layer several hundred meters above the ice surface.

TABLE 3.22. Average meteorological parameters for the period 00z 1 May through 12z 8 May 1968. The asterisk marks stations with averages only for the last six days, after arrival of the cold air.

	P_o(mb)	T(°C)	T_d(°C)	D_R	V_R(m/sec)	$\overline{V}$(m/sec)	q
a) Stations to the east and north of the peninsula							
MAT	994.6	-18.8	-21.6	226°	12.2	12.3	.99
PET*	994.6	-15.4	-17.8	176	17.3	18.4	.94
BHN*	992.2	- 8.6	-11.5	122	8.8	9.1	.96
b) Stations on the west side							
STO	990.5	- 7.9	-14.6	111	14.7	15.0	.98
ADE	998.5	- 5.9	-10.3	092	5.9	6.6	.89
FAR	998.4	- 4.4	- 8.6	091	1.5	3.5	.46
c) Stations south of 70°S							
BEL	1004.9	-26.8	-30.4	164	11.4	11.6	.98
HAB	1001.9	-21.3	-23.6	089	6.0	6.5	.93
FOB	992.2	-15.8	-20.1	324	1.8	2.4	.74
d) Upper air data for FAR							
FAR	850	-12.0	-17.7	164	1.2	5.4	.23
FAR	700	-21.8	-27.6	235	1.8	6.2	.30
FAR	500	-36.6	-43.1	225	3.8	8.2	.46

An examination of all three-hourly observations made in the twelve months March 1968 to February 1969 at MATIENZO and STONINGTON strongly suggests that there is a high correlation between the barrier winds along the peninsula's east side and the easterly foehn winds in Marguerite Bay. This relationship will be analyzed in the following Section.

3.3.5 Barrier and Foehn winds related

It is evident that synoptic observations on both sides of the mountains are needed to find out more about the relationship between barrier winds along the east flanks and foehn winds rushing down the west side escarpment. Fortunately, a few such observations were obtained in 1947 by the Ronne Antarctic Research Expedition, at Stonington Island in Marguerite Bay, 25 km farther east on the crest (CRE) at 1768 m above sea level, and at Cape Keeler (KEE) on the east side of the Peninsula (Fig. 3.12).

Summaries of the observations of these three stations have been published by the U.S. Navy (Peterson, 1948), but the synoptic data have received little attention. Copies of the original records of the American Expeditions of 1940-41 and 1947-48 were made available by the Center for Polar and Scientific Archives, Washington, D.C. The twice daily (00 and 12 GMT) observations and

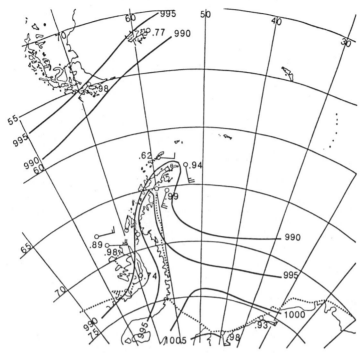

Fig. 3.16. Mean surface winds and sea level pressure conditions during an extended period of barrier winds on the east side of the peninsula and foehn in Marguerite Bay, 1-8 May 1968. One long barb of the wind arrows stands for 5 m/sec; the number is the directional constancy. Data have been plotted only for stations with high values of q. For the sake of clarity the data of the stations ESP, MAT, ADE, and SMA have been moved sideways.

Note the extremely high constancy of the winds (q) at MAT, STO, and BEL; for 3-8 of May it is very high also at PETREL and BELLINGSHAUSEN.

some additional notes of the observers contain valuable information about the character and origin of the strong katabatic winds which are a frequent phenomenon in Marguerite Bay, particularly in the vicinity of the steep rising mountains to the east, and about the simultaneous conditions on the eastern foot of the mountains as well as on the plateau. Even though the typical foehn events in Marguerite Bay appear more frequently and with greater force in fall and winter, the two months October and November include a fair sample of interesting cases.

The synoptic observations of all cases show quite clearly that during a foehn storm in Marguerite Bay the atmospheric pressure (at sea level) is considerably greater on the east side than on the west side of the mountains. This, of course, is not much news; it is known to happen in many other regions.

Of the ten cases of strongest easterly winds at STONINGTON Island (>15 m/sec at about 3 m above ground), nine had a pressure difference Δp (KEE - STO) > 10 mb, and in the tenth case the difference was 8 mb. Even allowing for possible dynamic effects of strong and gusty winds on the pressure inside a station building, the average for the ten cases (13 mb, σ_m = 1.1 mb) is large enough to support the above statement. Looking inversely at situations with particularly strong pressure differences, Table 3.23 gives the average values of wind and temperature for all 13 cases in which Δp (KEE-STO) > 10 mb, and for ten cases with Δp between 6 and 10 mb.

TABLE 3.23. Average surface wind (m/sec) and temperature values in cases with large positive pressure differences east side minus west side of the Antarctic Peninsula at 68.5°S, approx., October and November 1947. Symbols as in previous tables, and n = number of 00 and 12z observations; θ = potential temperature; p_0 = pressure at mean sea level. CRE = crest station.

a) Δp_0 (KEE-STO) > 10 mb; n = 13.　　　b) 10 mb > Δp_0 (KEE-STO) > 6 mb; n = 10

	a) STO (gm)	CRE (1768 m)	KEE (23 m)	b) STO	CRE	KEE
D_R	108°	051°	165°	087°	020°	179°
V_R	16.5	9.3	6.7	6.7	5.1	5.1
V	17.0	9.8	7.7	7.7	10.3	5.7
q	0.97	0.97	0.86	0.90	0.82	0.89
T	-3.9	-21.1	-14.6	-1.7	-16.7	-14.1°C
$\sigma(T)$	3.6	4.9	3.5	4.3	5.8	3.0°C
θ	272	272	260	273	275	260°K
p_0	969	760*	982	979	784*	986 mb

*computed, for station level

In the 13 cases summarized in Table 3.23a), the lapse rate of temperature between the crest at 1768 m and Stonington Island at 9 m is very nearly the dry adiabatic one, so that the mean temperature of that atmospheric layer can easily be determined. Since the horizontal distance between the two stations is only 25 km, the pressure at CRE can be computed with reasonable accuracy. A similar computation for the eastern side of the mountains is more problematic because the exact temperature profile between crest and Cape Keeler is not known and the distance between the two places is much larger (155 km). Still, there is no doubt that the air on the east side is considerably colder and the sea level pressure greater than on the warmer west side. Assuming temperature profiles characteristic for Weddell Sea air masses as suggested by kite soundings made during the drift of the DEUTSCHLAND in 1912 (Barkow, 1913; Schwerdtfeger, 1975) and HALLEY soundings on days with west winds, with a cold layer in the lowest 500 to 1000 m and warmer air above, one obtains a sea level pressure of about 980 mb at the foot of the mountains east of the crest station. For the 13 cases to which this consideration applies, the mean pressure at KEE, 85 km farther south, is 982 mb (computed from aneroid barometer measurements).

From these facts evolves a clear picture of the stream field. Winds with an east to west component over the Weddell Sea carry stable air masses toward the mountain wall of the peninsula. The cold air of the lowest layers piles up along the eastern slopes, so that the wind in the lower layers is deflected into a south to north direction, parallel to the mountain wall. In many cases this cold air does not reach the crest of the mountains. That is evident from the pronounced difference in potential temperature between KEE and CREST. Even a saturation-adiabatic rise of the air near the surface on the east side would not make much of a difference. It would lead to a potential temperature at the height of the crest station of 262°K (instead of 260° for a dry-adiabatic rise), to be compared with the 272°K determined for that station. In such situations, the weather along the east side of the peninsula is characterized by overcast sky, often light snowfall, and drifting or blowing snow whenever the strength of the surface winds is sufficient.

The weather conditions are quite different on the west side of the mountains. The typical phenomenon is a foehn storm which carries air from about crest height down into Marguerite Bay, as the potential temperature values for CRE and STO as well as verbal descriptions suggest. Note also the extremely high values of the directional constancy (q) of the wind in Tables 3.22 and 3.23. The moisture measurements at the two stations are probably not precise enough to justify a detailed analysis. For all 23 cases with Δp(KEE-STO) >6 mb combined, the relative humidity decreases from 78% (close to saturation with respect to ice) at the crest to 64% at Marguerite Bay, which corresponds to an increase of the specific humidity from .7 to 1.8 gram/kg. That could be attributed, at least in part, to the evaporation of blowing snow which is reported in most cases.

There is one case among the 13 summarized in Table 3.23a), 28 October 1947, in which the above mentioned conditions for the development of a foehn storm appear to be met, but nothing happens at Stonington Island itself. The pressure difference KEE-STO is 11 mb, KEE reports -14° with winds south-southeast 19 m/sec, CRE -13° with east 24, and STO +3°, but an east wind of only 4 knots. The observer writes: "Snow falling off the plateau in the east...Looks like someone is shoveling pile after pile of snow off the fringe. The splash is deflected by NE-Glacier." It may be that the non-development of a full grown foehn storm in this case is due to a lack of time; 12 hours before and after the critical time, Δp(KEE-STO) was less than 2 mb.

3.3.6 The winds on the western Ross Ice Shelf

Evidence of the mountain barrier wind effect in the western Ross Ice Shelf and Ross Island area is contained in the classical work of G.C. Simpson (1919) on the meteorological results of Scott's second expedition, 1910-13; Simpson also made use of observations of Shackleton 1907-09 and Amundsen 1911-12. In retrospect it can be said that the contrast between the relatively favorable conditions found by Amundsen on the eastern side of the Ross Ice Shelf, and the frequency of adverse wind and weather experienced by Scott on the west side in the summer 1911/12, fits well into the climatological picture of an area in which stable, cold air masses tend to move from the east toward a blocking mountain wall, in this case the Transantarctic Mountains and the high plateau of East Antarctica beyond.

The tendency of the stable cold air near the surface to follow the configuration of the terrain, moving parallel to the mountains and around the islands that block the way, was clearly recognized by Simpson, as proved by his sketch of the prevailing streamlines of the surface winds south of, and around, Ross Island during blizzard conditions (Fig. 3.17). The capital letters A through I denote positions for which evidence of the direction of the streamlines during blizzards was available, either from direct wind observations or from sastrugi formation on the snow surface. Here are only a few examples: a violent blizzard with south-southwest winds was observed at Cape Crozier, position G, on 22 and 23 July 1911 while at Cape Evans, position A, the wind was steadily from the east. "The winds at position I and to the south are clearly indicated by the sastrugi which all point to the high winds coming from the S or SSW." And, on 10 and 11 July 1911, "there was a blizzard at Cape Evans with winds from ESE, while the Cape Crozier party at position F experienced just as high and steady winds, but from the SW and SSW." Most remarkable, however, may be the following quotation from Simpson's work (1919, p. 111): "Position E. - Variable winds. On July 4, 1911, the Cape Crozier party was at the point indicated by E. Throughout this day an east-southeasterly blizzard blew at Cape Evans with velocities up to 52 miles an hour, but the party recorded: 'Overcast all day with steadily falling snow. Wind 3 to 4 (Beaufort) with occasional gusts from ENE to SE and light breeze from SSE', thus showing variable winds during a blizzard which was giving steady high winds from the ESE at Cape Evans. Farther within the large bay to the south of Ross Island the wind is always light and variable. The bay was explored right up to the coast by members of Captain Scott's first expedition and it was found to be full of light snow, showing little or no signs of heavy winds. E.A. Wilson called this bay Tranquil Bay." On modern maps it is called "Windless Bight".

The idea that the strong southerly winds parallel to the Transantarctic Mountains between 84 and 78°S must often be due to the damming up of air initially moving from east to west, a concept supported by the westward increasing snow accumulation (Fig. 5.5), cannot be found in Simpson's writings. But he is quite outspoken with regard to the frequency of these southerly winds, and to the fact that the prevailing easterly flow observed at Hut Point and the southeasterly at CAPE EVANS (the base station of Scott's second expedition) are a deflected branch of the airstream from the south. He reasons that the air moving through McMurdo Sound, between Ross Island and the mainland to the west, can be divided into two groups: one comprising the winds from the ice shelf generally northward to the western Ross Sea, the other in the opposite sense (Table 3.24). Thirty months of three-hourly wind data obtained by the automatic weather station MARBLE POINT (number 06 in Fig. 3.18) on the west side of McMurdo Sound confirm this concept (Table 3.25). To quote Simpson again (ℓ.c., p. 114): "The average velocity of the wind from the south was nearly twice that of the wind from the north. The total flow of air from the south was almost nine times that from the north". Of a total of 13,590 hourly wind data, only 1% indicates a flow from the north sector at more than 30 m.p.h. (about 14 m/sec),

TABLE 3.24. Relative frequency (%) of surface wind 6 to 15 and > 15 m/sec at CAPE EVANS, eastern shore of McMurdo Sound, coming from the Ross Ice Shelf and from the Ross Sea; computed from Simpson's (1919) table 65, period February 1911 - August 1912. Total number of hourly observations: 13,590; of winds > 5 m/sec: 7,924; of winds > 15 m/sec: 3,800.

air coming from Speed – Class	Ross Ice Shelf 6-15	> 15	Ross Sea 6-15	> 15 m/sec
Rel. frequency	40	46	12	2%
of > 15 m/sec only	--	96	--	4%

TABLE 3.25. Average wind speed and resultant wind vector (m/sec, measured 3 m above ground) and relative frequency of calms at AWS Marble Point, western shore of McMurdo Sound at 77.4°S, 121 m above sea level, for December and January 1980/81, 81/82 and 82/83, and for the cold season March to October 1980 and 1982. n = number of days.

Part of year	$\overline{V}$	V_R	D_R	q	calms (%)	n
Summer	3.6	1.3	152°	.37	5	169
Cold season	3.7	2.6	194	.70	17	488

but 27% of the total are such strong winds from the south. Direct observations made on the ice shelf prompted Scott to write in his diary, from his camp 13, southeast of Minna Bluff, February 1911: "All the sastrugi are from SSW to SW, and all the wind that we have experienced in this region, -- there cannot be a doubt that the wind sweeps up the coast at all seasons".

For his streamline sketch Simpson could already utilize the experience gained during Scott's first expedition 1901-04 and by Shackleton's 1907-09 (Shackleton, 1909). Curtis (1908, p.490 and table V) reported that at Hut Point (near the southwestern extremity of Ross Island) the most frequent wind direction all year round was from the east, but on the ice shelf south and southeast of the island, 56% of all 628 observations made on two sledge journeys between October 1902 and February 1903 were from the south or southwest, 27% calms, and only the remaining 17% winds from the other directions. Shackleton, going south from 78.6°S to the Beardmore glacier (10 November to 3 December 1908) about 40 km farther east than Scott did in 1911, also had to fight headwinds most of the time, but experienced only one blizzard day. From 82.3°S, 168°E he reports "abundant signs that the wind blows strongly from SSE during winter".

Up to here it appears evident that there are two processes which determine the occurrence of the persistent cold winds from the south sector. i: the piling up of stable boundary layer air initially moving from the east toward the impassable wall of the Transantarctic Mountains, and ii: much less frequently for sure, the presence of a more or less stationary cyclone over the central part of the Ross Ice Shelf or the southern part of the Ross Sea. It must be recognized, however, that there is still another type of atmospheric flow that can contribute to the frequency and strength of those winds in a band of a width between 100 and 200 km along the foot of the mountains. The transverse glacier valleys and troughs of the terrain between the individual higher elevations and peaks in the nearly 1000 km long mountain range framing the Ross Ice Shelf open ways not only for the ice masses lying and rising behind the range, but also for katabatic winds. Hence, whenever the horizontal pressure gradient at the shelf ice is weak, this fast moving air, after arrival at the foot not anymore guided by the configuration of the terrain, must turn to the left due to the Coriolis acceleration, as a kind of inertial flow. The flank of the Transantarctic Mountains between 78 and 84°S offers propitious conditions for this phenomenon to become visible. In the short summer, the adiabatically warming katabatic air often arrives in the lower layers with a higher potential temperature than that of the boundary layer over the ice shelf. Under such conditions the turning to the left of airstreams can be recognized in VHRR (very high resolution radiometer) infrared satellite pictures as shown by Swithinbank (1973). More about inertial flow patterns in Section 3.4.3.

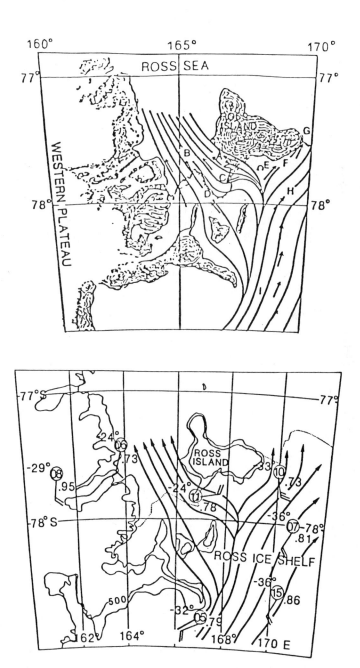

Above: Fig. 3.17. Simpson's sketch of streamlines near the surface during blizzards in the Ross Island region. The letters A to I are explained in the text.
Below: Fig. 3.18. Streamlines in the same region for all days of April 1982,
based on the continuous wind records of the automatic weather stations.

All the relevant meteorological facts measured or observed since the IGY confirm the hypothesis regarding the origin and character of the barrier winds along the wall of the Transantarctic Mountains (Schwerdtfeger, 1977). These winds are of the same type as those blowing along the east side of the Antarctic Peninsula. Among others, an important piece of information is the mean annual accumulation map shown in Fig. 5.5, one of the results of the Ross Ice Shelf Project (Clausen et al., 1979; personal communication Dr. Dansgaard, 1983). It leaves no doubt that rising motion must prevail in the lower tropo-sphere along the mountain barrier west and southwest of the ice shelf. From the eastern and central part of the shelf to the western, the annual accumulation increases from less than 10 cm water equivalent to more than 15 cm, near 83°S more than 20 cm. Further discussion in Chapter 5.

As far as the flow pattern is concerned, a comparison of Figs. 3.17 and 3.18 shows a striking similarity between the streamlines designed by Simpson, based on correct interpretation of sporadic observations made at ten different spots in 1911 and 1912, and the pattern resulting from the continuous records of several AWS in the early 1980s. No further comments on the charts seem to be needed, but a related problem that can now be attacked with the help of automatic stations has to be mentioned. It is the question of the width, and eastward decrease of strength, of the mountain-parallel cold air stream called barrier wind. This is of particular interest for the wind-driven ice transport wherever such winds blow over large sea ice fields as found all year round in the western Weddell Sea, in winter and spring also in other regions. The afore mentioned theoretical study by Parish (1983) suggests that 200 km width of the barrier air stream might be an appropriate estimate.

Close to Ross Island, the records of the automatic weather stations indicate that Simpson correctly sketched what happens to the cold and stable air masses of the boundary layer when they approach the island. The northward moving air suddenly confronts a new obstacle; the terrain of more than 1600 m elevation extends over 60 km from west to east, crowned by Mt. Erebus (3800 m) and Mt. Terror (3200 m). Under these circumstances there is an array of surface winds that cannot be explained by the same simple reasoning as was applied to the larger-scale barrier winds along hundreds of kilometers of mountain walls. Winds from the northeast blow along the east side of the Hut Point Peninsula, the southwestern spur (reaching about 300 m elevation) of Ross Island. New Zealand's station SCOTT BASE lies in this stream. For the years 1968 - 1977, the resultant wind values computed from 85840 hourly observations are: direc-tion = 035°, magnitude = 3.0 m/sec, constancy = 0.57. Such a flow is opposite in direction to the wind vector that could be directly derived from the vector scheme shown in Fig. 3.15. An explanation might be based on the fact that streams of stable cold air from the south must pile up on the southern escarpment of central Ross Island to a height level far beyond the height of the Hut Point ridge. This could lead to a mass distribution and hence a horizontal pressure field that makes the observed winds possible. At present, however, that cannot be more than a vague hypothesis. Finally, there is another characteristic of the winds at SCOTT BASE. The prevalence of winds from the northeast is greatest when their speed is less than 10 m/sec. For winds greater than 15 m/sec the situation is completely different. In 80% of all such hourly observations (1478 in the ten years 1968 - 1977), their direction is from the south. These findings require further investigation which now, after the deployment of several automatic weather stations in the Ross Island region, appears well feasible.

3.3.7 Minor barrier effects

The shelf ice or fast ice surfaces along the east flanks of the Antarctic Peninsula and the Transantarctic Mountains are the two regions where the great vertical stability of the boundary layer air and the topography combine to produce the most favorable conditions for the development of barrier winds. These mountain-parallel streams of cold air can extend up to 1000 km. Over large areas of open water, the temperature inversion in the boundary layer gradually disappears or, at least, loses strength and can persist only at a level several hundred meters above surface. Under such conditions, a weaker barrier wind will develop, if at all. The main point is that a decrease of stability in the boundary layer diminishes the chances for said winds.

This line of reasoning implies that the inverse statement must also be true. For instance: northerly or northwesterly winds can carry comparatively warm maritime air from the South Orkney - South Sandwich region toward the northeast coast of the Weddell Sea (about the 700 km stretch between HALLEY and Cape Norvegia, 71.4°S, 12.3°W). That means the air moves over colder and colder water, then pack ice, and finally the Riiser L. and Brunt Ice Shelves with the western slopes of East Antarctica in the background. There is no doubt that this way northeasterly barrier winds can develop. As in Section 3.3.3, it shall not be said that all strong northeasterlies at HALLEY are of the same type, but rather that the barrier effect contributes to the frequency and strength of winds from that direction. That is supported by wind statistics for the years 1956 - 1964: of all three-hourly surface winds >12 m/sec (≥ Beaufort 6 according to the old scale used at HALLEY in those years), i.e., 410 cases of a total of 2130 observations for July, 1956-64, 61% were from the 80-100° sector, 34% from the 50-70° sector, and only 5% from all other directions.

Similar conditions can exist, but less frequently and only in the winter and early spring, on the west side of the Antarctic Peninsula when westerly winds blow over a broad belt of ice in the northeastern part of the Bellingshausen Sea.

3.4 OTHER REMARKABLE TYPES OF SURFACE WIND

3.4.1 A foehn storm at MATIENZO

Attention was directed in Section 3.2 to the type of katabatic wind that originates when very cold and stable boundary layer air is available at high elevation and finds its way downslope, more or less, to the coastal regions. It occurs in the most pronounced form in East Antarctica and presumably also in poorly known parts of the coast of West Antarctica between 80 and 140°W. Depending on the kind of air mass in the boundary layer near the coast -- relatively warm and moist air advected from the southern ocean, or stagnant air cooled mainly by radiative energy loss, -- the onset of the katabatic wind near sea level will be accompanied by either a decrease of temperature (the "Bora" type) or an increase of temperature by a few degrees (the "Foehn" type), two names much in use in European meteorology (Barry, 1981). A different kind of katabatic winds was described in Section 3.3.5, foehn storms that appear frequently in Marguerite Bay (M.B. in Fig. 3.12) when cold air piles up on the other side of the peninsula. Their peculiarity is that air masses carried across the mountains come from the free atmosphere, presumably the layer between 500 and 1500 m above the surface of the ice-covered Weddell Sea. The wind measurements on the crest of the peninsula and a comparison of the potential temperatures leave no doubt of this being so. These "new" air masses can reach the lowest layers over Marguerite Bay only when the sea level pressure field in the southeastern corner of the Bellingshausen Sea makes the outflow of the "old" air possible.

Another type of katabatic winds can be found in the region of the northern half of the Antarctic Peninsula. Whenever the low pressure (sea level isobars) trough between mid-latitude westerlies and polar easterlies over the Bellingshausen Sea is situated far enough south, relatively warm maritime air with neutral or only slightly stable vertical structure can move from the northwest quadrant toward the peninsula. Under such circumstances it is well possible that air of the lower layers rises on the upwind side and moves over the mountain range. Again, only when the sea level pressure field on the east side favors a southeastward transport of the cold boundary layer air, katabatic motion of the warmer maritime air downward to the lowest layer is feasible. The development of such a foehn storm, analysed by Schwerdtfeger and Amaturo (1979), is shown in Figs. 3.19 and 3.20. These winds are similar to the southern foehn of the European Alps, the region where the name foehn had its origin.

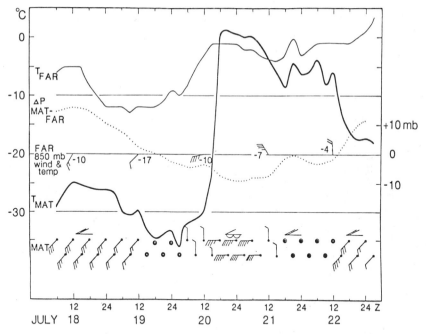

Fig. 3.19   Foehn storm at MATIENZO, 20/21 July 1968.
Thin solid line in upper part = temperature at FARADAY; heavy solid line = same
at MATIENZO, scale on left side ordinate; dotted line = sea level pressure difference MAT - FAR; wind arrows and temperatures along the horizontal center
line refer to the 850 mb (1300 m) level at FAR; 1 barb = 5 m/sec; wind arrows,
small circles for calm, and cloud symbols in the lower part are surface observations at MAT.

Two days before the onset of the phenomenon, the sea level pressure on
the east side was about ten mb higher than on the west side of the mountains,
and the typical cold southwesterly winds were blowing over the northwestern
Weddell Sea. The following day (19 July) brought decreasing winds, later calms,
clear sky, and the east to west pressure gradient reversing. The first half of
the 20th brought a very weak northerly flow with increasing temperatures, until
at 17 z (13 hrs local time) a westerly foehn storm broke through to the surface
with a spectacular rise of temperature, from -23° at 15 z to + 1 at 18 z. The
minimum on the 19th between 18 and 22 z was -38.2°, the maximum 24 hours later,
+2.2°. The storm at MATIENZO persisted for 16 hours, followed by a short period
of weak northerlies and almost 24 hours of calm, until the "normal" pattern,
cold and strong southwesterlies, imposed itself again. Temperature increases of
24° in three hours, or 40° in 24 hours, are certainly rare events anywhere, but
not unique. A case with a similarly large temperature change on the east coast
of Greenland at 75°N was described by Schatz (1951).

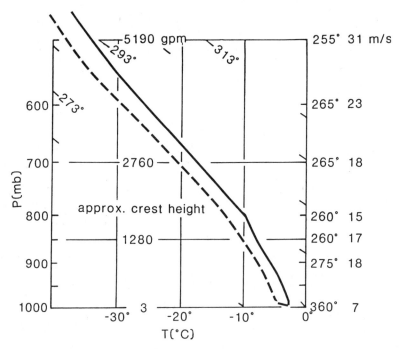

Fig. 3.20. Temperature (solid line) and dew point temperature over FARADAY, 12 z 20 July 1968, a few hours before the onset of a foehn storm at MATIENZO on the other side of the Peninsula. Along the right ordinate: wind at FARADAY at various pressure levels.

The surface winds on the other side of the mountains, at FARADAY, remained rather dull all the five days to which Fig. 3.19 refers; in the three-hourly synoptic observations, no speeds of more than 8 m/sec appeared. During the foehn event the station reported moderate winds from the northwest-sector, interrupted around midnight by weak northeasterlies. Nevertheless, the sea level pressure difference FAR-ADE (mean latitude 66.5°S) amounts to 15 mb, corresponding to the geostrophic wind component of 25 m/sec from the west-northwest. Furthermore, the sounding of FARADAY made at 12 z, only a few hours before the foehn event at MATIENZO, shows that at only a few hundred meters above the surface a much stronger wind must have been blowing toward the mountain range (Fig. 3.20). Indeed, all available data strongly suggest that air from the lower layers at FARADAY passed across the mountains and reached the station MATIENZO in a true katabatic motion on the lee side, after an anabatic motion on the upwind side of the divide. The humidity measurements near the surface at MATIENZO for the first 12 hours of the storm indicated a mixing ratio of 2.6 gram/kg. In the 12 z sounding at FARADAY, the same value belonged to the layer 200 to 400 m above sea level. Since routine measurements of the moisture content of the air at temperatures near and below the freezing point can generally not claim a high degree of exactness, it is worthwhile to note that the time series of the potential temperature values supports the above conclusion regarding the level of origin of the foehn air. During the foehn storm the potential temperature at MATIENZO was only 2° higher than the surface value at FARADAY, about 8° lower than the value which the sounding indicates for the level of the crest height.

At the station ESPERANZA, 230 km northeast of MATIENZO and 25 km south of the northern tip of the peninsula, the foehn event appeared in similar form, with winds from 260°, but speeds not more than 16 m/sec. It began about three hours earlier than at MATIENZO with a temperature increase from -16 to 0°, reached a maximum of +2° and persisted for nearly 23 hours. The relative humidity during the foehn varied at both stations between 55 and 65%, well compatible with the conditions on the other side of the mountains, though quite high in comparison to well known foehn events at the foot of the Andes or the Rocky Mountains. Some examples of foehn storms at Hope Bay (Table A1, note 3) were already given by Pepper (1954). Table 3.20 (p. 87) indicates how often typical foehn winds have been observed at MATIENZO in the twelve months period March 1968 to February 1969. It must be noted that at the latitude of 65°S the frequency of occurrence of strong westerly or northwesterly winds in the free atmosphere, above the crest height of the peninsula, increases with height. However, on many days with such upper level winds the sea level pressure field on the east side of the mountains does not force the cold air in the lowest few hundred meters to move away from the mountains.

For the five months June to October 1968, the frequency and duration of foehn storms at station ESPERANZA have been compared with the values of MATIENZO. The differences are not significant. South of 65°S latitude, where the number of days with strong westerly flow on the west side of the peninsula is smaller, foehn storms on the east side of the mountains must be expected to be less frequent; but there is clear evidence that they occur occasionally (NASA, 1971; Sissala et al., 1972; Gloersen et al., 1973).

The end of this Section might be the right place to mention one of the curiosities of antarctic weather brought about by the foehn effect. During the Swedish South Polar Expedition 1901-1903, regular meteorological measurements were carried out at SNOW HILL for 20 months (Bodman, 1910). The lowest temperature for the entire time interval was recorded on 6 August 1902: -41.4°C. The highest temperature (including the summer months) on 5 August 1903: +9.3°C.

3.4.2 Foehn and barrier winds favoring the drift of ice

According to reports of various kinds of expeditions which in the last 150 years have tried to go south by ship in the western Weddell Sea, any progress beyond the fittingly called Erebus and Terror Gulf (about 64°S, 56°W, Fig. 3.12) almost always was impeded by masses of pack ice; or at least, it was considered, with good reason, highly inadvisable to trust an occasional lead. In 1893, nevertheless, Captain C. Larsen, sent out by a Hamburg whaling company, reached much higher latitudes.   To quote Priestley et al. (1964),   Larsen "discovered land, afterwards named the Oscar II and Foyn Coasts, and came as far south as 68°10' in the western Weddell Sea."   This appears in hindsight as an exceptionally fortuitous enterprise, particularly since in recent years satellite pictures have shown (NASA, 1971; Sissala et al., 1972; Colvill, 1977) that occasionally large leads of ice-free water off the Larsen Ice Shelf can open and close within a few days. In extreme cases, these leads have reached a width of 50 km and extended from 65 to about 70°S.

A close relationship between such alternations of the ice and the synoptic weather situations can be shown to exist (Kyle and Schwerdtfeger, 1974). For six cases of opening and five cases of closing found east of the edge of the Larsen Ice Shelf on the daily, high quality DMSP satellite pictures of 51 fall and 70 spring days in 1973, the trajectories of cyclonic storms passing eastward through the peninsula region were determined from the twice daily synoptic analyses of the Australian Bureau of Meteorology. Fig. 3.21 summarizes the results: When there are strong horizontal gradients in the sea level atmospheric pressure field, the movement of the pack ice at sufficient distance from the peninsula is approximately in the direction of the geostrophic wind. This relationship between the sea level atmospheric pressure field and the drift of the pack ice is well known since Nansen's historic journey through the Arctic Ocean (Nansen, 1897); the theory was developed by Ekman (1902). The following conclusions can be drawn: Whenever a storm center comes to lie between the longitudes 65 and 50°W, forceful winds blow over the western Weddell Sea. These winds must be from the western sector only if the storm center is located south of the affected region. The much more frequent situation, over the years, is the opposite, with the cyclone centers' trajectories north of 65°S and easterly winds over the northern Weddell Sea, transforming into coast-parallel barrier winds when approaching the peninsula. (It must be noted that between 69 and 65°S the distance from the edge of the Larsen Ice Shelf to the eastern foot of the Peninsula mountains varies from 100 to 200 km). Naturally, the smaller frequency of the westerly wind situations explains that the western half of the Weddell Sea, in contrast to the eastern side, is generally covered

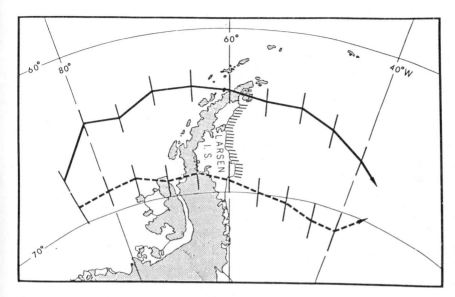

Fig. 3.21. Averaged cyclonic storm tracks computed from 40 individual trajectories in six episodes of opening leads, and 41 trajectories in five episodes
of closing leads, east of the Larsen Ice Shelf. The length of the meridional
line segments crossing the trajectories every five degrees of longitude indicate ± one standard deviation of the mean latitude of the cyclonic centers' locations. Twice daily synoptic analyses of the Australian Bureau of Meteorology,
supplemented by daily DMSP satellite pictures, for the periods 28 Feb.- 20 Apr.
and 15 Sep. - 24 Nov. 1973 have been used.

by 9/10 or more ice. Nevertheless, the occasional occurrence of strong westerlies or northwesterlies, needed to produce foehnstorms on the east side of the
Antarctic Peninsula, can contribute to make the northeastward drift of ice more
efficient than the barrier winds alone could bring about.

3.4.3 Appearance of inertial flow patterns

In parts of the coastal regions of the Antarctic (and the Arctic) it is
possible that a fast, more or less continuous stream of cold air suddenly
intrudes a new environment where the horizontal pressure field is entirely unrelated to the stream's original direction and speed. What then happens must
depend mainly on the horizontal pressure gradient in the receiving area.   When
this gradient and the surface wind in the preceding time are weak or there even
are calm conditions, it appears that the cold air stream can maintain its identity for some length of time, subject to the type of surface, ice or water,
over which it is moving. Then the Coriolis force (proportional to the wind
speed) cannot be in balance with a weak pressure gradient force, and acts to
gradually deflect the airstream to the left, creating what is known as inertial
flow.

Situations of that type can frequently be observed in the region of the Bransfield Strait, i.e., the waters between the northeastern part of the Antarctic Peninsula and the South Shetland Islands.   A branch of the strong, cold barrier winds along the west coast of the Weddell Sea often rushes through the Antarctic Sound.   The latter is 20 to 30 km wide (between the end of the Peninsula and the mountainous Joinville Island), and ill reputed for its dangerous gales and corresponding state of the sea. At the northern exit of the Sound, the stream of cold air suddenly arrives with all its momentum in the Bransfield Strait where no barrier effect can be expected (Parish and Schwerdtfeger, 1977). Fig. 3.22 shows such a situation, persisting without significant changes for 24 hours, at least. The strong south-southwest wind in the Antarctic Sound was recorded at the two neighboring stations ESPERANZA and HOPE BAY situated in a minor embayment of the west side of the Sound.   180 km to the

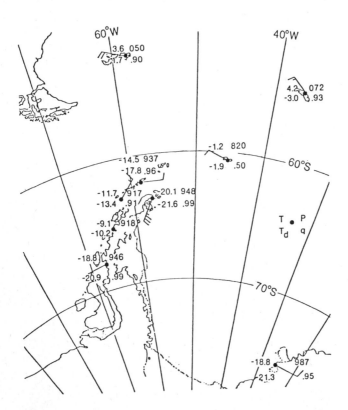

Fig. 3.22. Average weather conditions for nine consecutive three-hourly observations, 06z 4 to 06z 5 August 1959. The two numbers to the left give the air temperature and dew-point temperature, to the right sea level pressure and directional constancy of the wind.

north-northwest, at Station ADMIRALTY BAY (on King George Island), this stream of cold air has diminished in speed and turned to the left. At the beginning of this episode in the night 3-4 August, the south-southwest wind at HOPE BAY increased from 2 to 15 m/sec between 00z and 06z; at ADMIRALTY BAY the wind changed from north 10 to east 15 m/sec and the temperature from 0° to -14°C with a delay of three to six hours. Similar though somewhat smaller variations appear at DECEPTION Island some 200 km downwind, with about six hours delay against ADMIRALTY BAY. It is typical for such situations that temperature and moisture content of the air at the South Shetland Islands is considerably lower than at station FARADAY, 250 km SSW of DECEPTION Island. Similar cases were already observed by G. Robin (unpublished Base report and personal communication) and, earlier still, in the first years of this century (Meinardus and Mecking, 1911). Only the inertial flow hypothesis is less than ten years old.

Assuming steady state, no horizontal pressure gradient at all, and the friction force proportional and directed opposite to the velocity of the wind, integration of the equation of horizontal motion (in natural coordinates) leads to the equation

$$V = V_0 \exp(-\frac{t}{t^*}) \tag{3.7}$$

where V = velocity as function of time t, V_0 = initial velocity and t^* = frictional time-constant ranging from 10^5 sec (less efficient friction) to 10^4 sec (more). Fig. 3.23 shows trajectories across the Bransfield Strait for V_0 = 20 m/sec (Parish, 1977). Admittedly, this kind of trajectory computation is oversimplified, and modern theoreticians will dislike it. Nevertheless, it seems to suffice to show the main line of thought. A more rigorous approach would take into account that the cold air low level jet, intruding into the air over the Bransfield Strait, modifies the initial pressure field so that the theory of geostrophic adjustment should be applied, and so on. That shall here not be followed up. The main point is that the scale of inertial flow under the described conditions is in accord with the observations.

On a larger scale, an inertial flow effect can be important for the low-level atmospheric (and thus also oceanic) circulation, in particular in the east antarctic off-shore regions between about 160°E and 10°W. The considerations shall again refer to synoptic situations in which the horizontal pressure gradient over the waters or pack ice fields near the coast is weak. Differing from the model shown in Fig. 3.23, however, there can be strong off-shore streams of cold air across a coastline of hundreds of kilometers. It appears possible that the cold air masses, initially from the south or southeast, can maintain their identity for some time and increase the east to west component of their motion significantly.

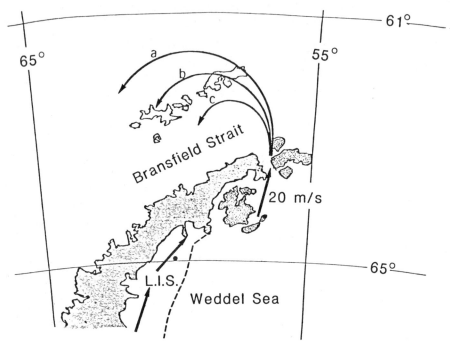

Fig. 3.23. Trajectories of a low-level jet stream of cold air in inertial motion after passing through the Antarctic Sound. Assumed initial speed is 20 m/sec. Trajectory a for frictionless flow, b and c for flow modified by weaker and stronger frictional effects.

The strong and frequent off shore stream in the east antarctic coastal regions is, of course, the katabatic wind, as it is well documented for a fair number of stations, and described in detail in Section 3.2. Unfortunately, there is little comparable meteorological evidence to show what is going on over the coastal waters, almost no measurements at all for the months in which the katabatic winds are strongest. Table 3.26 contains an inhomogeneous collection of surface wind data as function of distance from the coast in the Davis Sea. In spite of observational shortcomings, it is almost inevitable to conclude that the east antarctic katabatic winds, created mainly by radiative cooling of the air at high elevation and the appropriate topography to guide the air down to the coast, act to reinforce and maintain the circumpolar easterlies near sea level. That would also be in agreement with the well known fact that these easterlies are stronger and much more persistent than the north polar easterlies which do not get such a well organized support. How much this support also affects cyclogenesis in the high southern latitudes and the position of the circumpolar low pressure trough, still remains an unanswered question.

TABLE 3.26. Mean speed and prevailing direction of surface winds at MIRNY and offshore between 86° and 98°E, as function of distance northward from coast.

Gauss = ship of the German Antarctic Expedition 1901-1904, blocked by the ice at 66.0°S, 89.6°E, Mar. 1902 - Feb. 1903;

Grottoes = west station of Mawson's Australasian Expedition, 1911-1914, on the Shackleton Ice Shelf at 66.3°S, 95.0°E, Apr. 1912 - Feb. 1913;

Other names = temporary Russian stations; MIR was located on the ice dome of Drygalski Island, about 330 m a.s.l., reference: Tauber, 1960; Skeib, 1963; Mather and Miller, 1967.

Name	Distance from coast, km	Surface	Mean Speed m/sec	Prevailing Direction	Apr. - Sep. when only years given
MIRNY	- 1	near shore	12.5	SSE	1957-73
MS 4	+ 14	fast ice	4.4	SE	August 1956
GROTTOES	27	ice shelf	5.6	ESE	1912
DRUZNAYA	40	ice shelf	7.7	E	May-July 1960
MIR	80	Dryg. Isl.	10.1	E	May-July 1960
GAUSS	85	fast ice	7.3	E	1902
POBEDA	250	ice shelf	7.2	E	May-July 1960

When the horizontal pressure gradient and correspondingly also the surface winds over the open sea are strong, intense mixing of two very different air masses, cold continental versus milder maritime, must occur and the individual characteristics of the relatively shallow cold air will soon be blotted out. Some evidence to support this notion was given by J.K. Davis, captain of the expedition's ship, in the famous book "The Home of the Blizzard" by Mawson (1915). Davis describes his observations made while at anchor in the Commonwealth Bay (near CAPE DENISON) during a short "calm" period, in the following words: "To the north, violent gusts appeared to be travelling in various directions, but to our astonishment these gusts, after approaching our position at a great rate, appeared to curve upwards; the water close to the ship was disturbed, and nothing else. This curious phenomenon lasted for about an hour; then the wind came with a rush from the southeast, testing the anchor chains in the more furious squalls". Obviously, no traces of inertial flow can be identified under such conditions.

3.4.4 Crossing of air streams on the Adelaide Piedmont

An answer to the question whether, or for how long a time, two different currents of air in the boundary layer, moving at nearly a right angle one to the other, can maintain their identity, would be of interest for the problems discussed in the previous section. It probably happens quite seldom that actual observations can be made at a place and time when a cold, gravity-driven air stream is undercutting a larger size stream responding to the horizontal pressure gradient. That can occur at the west side of Adelaide Island (67.5°S, 69°W) where strong winds from northeast or north-northeast characterize the prevailing wind field near the surface. At the station of the same name, located at the southern tip of the island and in operation from 1962 to 1975, strong winds are generally from north or northwest; the stations's site is well sheltered from the northeasterlies (direction between 20 and 70°) by the island's mountains. In order to investigate the change of direction of fast moving air masses over a relatively short distance, meteorologists working at ADELAIDE ISLAND in the 1960s repeatedly travelled to Lincoln Nunatak, an outstanding feature of the terrain 40 km north of the station. The team of May 1967 (25 to 29) encountered particularly interesting conditions, reported by F.W.A. Wilkinson (Preliminary Report 1966-67, not published, used here with permission of the Director, British Antarctic Survey).

At the day of departure, winds from the N were recorded at the station itself. About 10 km to the north (Fig. 3.24), the surface wind was blowing from the east, off the glaciers and snow-covered mountains. Ordinary air-filled balloons were released and tracked with a compass till they were out of sight; all readings indicated a flow from 85°. Only in the neighborhood of Lincoln Nunatak and farther northeast, winds from northeast were observed. The easterly winds were carrying and depositing snow. They must have continued or repeated themselves a few times in the following days because on the return journey large sastrugi were observed, formed in east to west paths of the winds off the glaciers. It seems that the strong winds over the western half of Adelaide Island are from the northeast, nearly parallel to coast and mountains, until they reach the latitude of Lincoln Nunatak; later observations of sastrugi and snow dunes in the northern part of the island back this up. South of Lincoln Nunatak, where the mountains to the east are higher and wider, the northeast winds rise over the stronger and somewhat colder glacier winds, continue southward, and re-approach the surface where the glacier winds cease, about 10 km north of the station. It has also been observed that the easterly winds blow drift-snow out to sea, over the ice cliffs; several miles out, the drift takes a more northeast to southwest path (Fig. 3.24).

3.4.5 Bimodal wind regimes

The Wright Valley is a west to east trough, 46 km long and 8 km wide. The station VANDA is located near the eastern end of Lake Vanda, at the lowest point in the valley, 75 m above sea level, about 2000 m below the eastern rim of the east antarctic plateau (Thompson et al., 1971; Riordan, 1975). The lake,

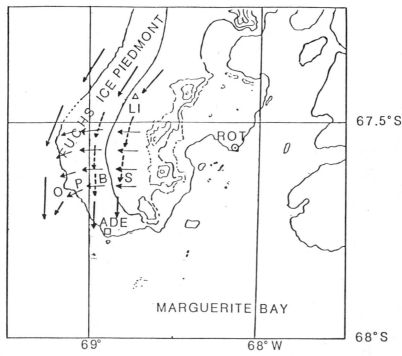

Fig. 3.24. Crossing of airstreams over the Adelaide Piedmont, observed in May 1967. Solid arrows = winds near the surface; dashed arrows = winds a few tens of meters (estimated) above. Contour lines of the terrain at 500 m intervals; highest peaks > 2000 m. ADE = Adelaide Island station (1962 - 1975); ROT = Rothera station; LI = Lincoln Nunatak. O.P.B.S. = observed path of blowing snow.

the valley, and the surrounding mountains have been investigated repeatedly since the IGY 1957-58, though complete meteorological records are available only for the three years 1969, 1970 and 1974.

VANDA is one of the rare places in the Antarctic where the mean wind speed of the four summer months (November to February: $\overline{V}$ = 6.6 m/sec, $\overline{T}$ = -2°C) is almost twice as large as the winter mean (April to September: $\overline{V}$ = 3.5 m/sec, $\overline{T}$ = -33°C). The narrow valley does not leave much room for north or south winds to impose themselves. Fig. 3.25 shows a typical bimodal frequency distribution of the wind direction. Thompson et al. (ℓ.c.) observed that "a well-developed sequence of up- and down-valley winds blows in summer, with easterlies of 8-10 m/sec during the warmer part of the day and westerlies of 8-10 m/sec "overnight", when the input of solar radiation is at its minimum. Westerly winds were more prevalent in the winter and spring of 1970, but they were related to strong southerly to westerly winds aloft rather than to katabatic effects."

This statement must also apply to the few cases in the winter night of 1969 in which a period of calm and slowly decreasing temperature was suddenly

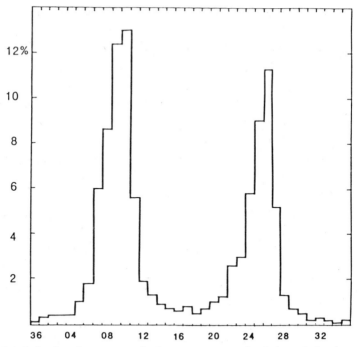

Fig. 3.25. Relative frequency of the occurrence of winds > 1.5 m/sec at VANDA, as function of direction; the latter is marked along the abscissa in tens of degrees. Source: Thompson et al., 1971.

brought to an end by rather strong westerly winds and an abrupt rise of temperature near the surface from lower than -40 to higher than -20°C, without the appearance of clouds (Riordan, 1975). As the winter temperatures of VOSTOK and DOME C indicate, it is not likely that such warm air came as katabatic wind from the plateau to the west of the dry valley. Also, the very cold and stable air existing in the boundary layer in the valley prior to a sudden warming cannot easily be removed through motions imposed by a mesoscale horizontal pressure gradient, because the valley is surrounded by mountains on all sides. It must therefore be concluded that the sudden warmings occur when the circulation within the confines of the valley happen to lead to divergent flow; that, of course, can only occur locally, not simultaneously over the total area. The short duration of the warming spells shown by Riordan (ℓ.c.) and the different surface wind patterns at his four auxiliary stations a few kilometer distant from VANDA support this notion. If that is so, the warm air above the surface inversion can subside and approach the surface.

At the northeast corner of East Antarctica lies CAPE ADARE, the tip of a narrow, about 50 km long, mountainous peninsula jutting out northward from Victoria Land's Admiralty Mountains. It is a well known spot in the history of antarctic exploration as the place where for the first time meteorological and magnetic measurements and observations all through the winter (1899) were made on the continent itself. When the Norwegian C.E. Borchgrevink, the leader of the "Southern Cross" Expedition, and L. Bernacchi (1903) published the results of their endeavor, they made the same bitter experiences with the wind records as Mawson and Madigan did with the CAPE DENISON winds some twelve years later: the professionals at home did not want to believe it. Rather soon, however, Borchgrevink was vindicated by the reports of the "Terra Nova" Expedition's northern group which stayed at CAPE ADARE March to December 1911 (Priestley, 1914). It was true. The frequency of calms at the station is amazingly large, but vehement windstorms greater than 30 m/sec also appear occasionally, in fact ten such storms between May and August 1911. That leaves little time for moderate winds. Here a few numbers from Simpson (1919) and Meinardus (1938): 37% of all 217 observations in May indicated calm, 42% winds or gales from the SE quadrant with a mean speed of 16 m/sec, and 21% winds from the three other quadrants, with a mean speed of 2 m/sec! For the complete series March to December 1911, calms or winds < 3 m/sec (force 1 or 2 of the Beaufort scale) were reported in 79% of all 2600 observations.

Simpson (ℓ.c.) called the wind conditions at CAPE ADARE one of the paradoxes of the Antarctic. It has to be considered, however, that i: the station was located on the west side of the peninsula about 2 km south of the end of the promontory itself, ii: the mountain ridge east of the station reaches more than 300 m above sea level and a few kilometers farther south rises to more than 1200 m, and iii: easterly and southeasterly winds are the most frequent surface winds over the Ross Sea. These circumstances make the high frequency of calms and insignificantly weak winds quite understandable. The strongest storms from the east-southeast were observed when the pressure was lowest and the temperature far above normal. So it can be concluded that the easterlies not only pass the ridge of the peninsula, but also descend on its lee side immediately and forcefully, provided an intense cyclone happens to be situated a few hundred kilometers north of the cape. The point is that under such conditions the pressure gradient over the waters to the west of the station brings about a northwestward directed airflow, out of the Robertson Bay. That should lead to weather as described in the first episode of Table 3.27. When the indicated condition is not fulfilled, the steep ridge of the peninsula can shelter a place close to the western escarpment quite efficiently.

TABLE 3.27.   A stormy and a quiet period at Cape Adare, May 1911
              (from Meinardus, 1938).

Date	8 - 12 May		26 - 29 May	
Wind average	SE	25 m/sec	Calm or NW < 1 m/sec	
max	SE	35 " "	NNW	2.5 " "
Temp. average	- 7.9°		-30.1°	
max	- 4.4		-26.7	
min	-11.1		-33.3	
Pressure ave.	982 mb		989 mb	
Cloudiness	9.2 tenths		1.2 tenths	
Observer's remarks	incredible gusts		- -	

3.5 BLOWING SNOW

In meteorological terminology there is a difference between "drifting snow" and "blowing snow". The first term is used when masses of fine snow particles are carried by the wind only in the lowest layer above the surface, so that the horizontal visibility at the height of the observer's head is not impaired. Depending upon the turbulence in the lowest layer, the smoothness of the surface itself, and the availability of easily transportable material, drifting snow might be initiated and maintained by wind speeds anywhere between 5 and 10 m/sec, becoming more dense and increasing its vertical extent as wind and its turbulence intensify. Altogether, a phenomenon well known to any attentive observer in mid latitude continental climates.

The second term, blowing snow, describes the much more vehement type of wind-driven snow. The vertical extent of the snow particle-filled atmospheric layer may be between 2 m and, under extreme conditions, a couple of hundred meters above, and the horizontal visibility strongly reduced, in extreme cases to less than 10 m. At PORT MARTIN, one of the two windiest places known in the Antarctic, visibility less than 10 m happened at 96 of a total of 2823 six-hourly observations in the 24 months February 1950-January 1952, corresponding to 3.4%; in the 14 months March-September 1950 and 1951 it was 5.5%. Visibility less than 1 km was recorded in a 34% of all six-hourly observations of the two years, 45% in the 14 months named (from Le Quinio, 1956). Fritz Loewe, who spent the hard year 1951 at PORT MARTIN, confirmed: "These low visibilities are almost exclusively caused by blowing snow" (Loewe, 1970, 1972, 1974). Therefore, short visibility statistics might be of interest: Table 3.28.

Naturally, it is often impossible for the observer to decide whether there is also precipitating snow in the air, in addition to the true blowing snow which has been lifted from the surface. As neutral term including both possibilities, the name blizzard might be appropriate, and is frequently used.

TABLE 3.28. Relative frequency (%) of different ranges of visibility at PORT MARTIN, Feb. 1950 - Jan. 1952.

Range		NOV - FEB	MAR - OCT	YEAR
< 50	m	3	19	14%
50 - 500	"	7	19	16
500 - 2000	"	2	9	7
2 - 10	km	11	14	13
> 10	km	77	39	50

When one wants to get a rough estimate of the frequency or climatological probability of occurrence of drifting snow of moderate intensity and of blowing snow which practically inhibits any out door work and travelling, frequency statistics of wind speeds greater than 10 and greater than 20 m/sec can be helpful. Bear in mind, though, that strong winds should be related to visibility-decrease due to blowing snow only if snow at the surface, upwind, can be assumed to be abundantly available, as it is in the first days after a real snowfall. After several days with light winds and no snowfall, a thin crust can form and restrain the drifting. Notwithstanding, the strong wind frequencies in Table 3.29 show the contrast between differently exposed stations; for instance, the chances for major blowing snow events at the South Pole are rather small.

TABLE 3.29. Relative frequency (%) of wind speed 11-20 and > 20 m/sec. The values for the upper eight stations have been computed from Phillpot (1967); for PORT MARTIN, only the annual values for 1951 have been published (Le Quinio, 1956).

Station	NOV - FEB		APR - SEP		YEAR	
	11-20	> 20	11-20	> 20	11-20	> 20 m/sec
SOUTH POLE	2	0	14	0	10	< 0.01 %
LIT. AMERICA	6	< 0.1	15	< 0.4	12	< 0.3
ELLSWORTH	7	< 0.3	24	< 0.5	18	0.3
MCMURDO	12	< 0.1	25	1	22	0.5
BYRD	22	0.2	37	2	32	1
WILKES	10	2	14	5	12	4
MIRNY	32	2	55	12	48	7
MAWSON	43	5	48	12	45	9
PORT MARTIN	N.A.		N.A.		29	44

The results of a notable research project on snow drift carried through at BYRD Station in 1962 and 1963, including theoretical considerations regarding drift content and total drift transport, have been published by Budd, Dingle, and Radok (1966). Only a few important points of this extensive work, with numerous references to earlier studies of wind-driven snow, shall here be mentioned.

i: The drift density (gram m^{-3}) as function of height at different wind speeds V (m/sec). Measurements of vertical wind profiles and collection of drifting snow in so-called "traps" lead to values as shown in Table 3.30. Introducing the visibility as function of drift density according to Liljequist (1957) and Budd et al. (1966), the horizontal visibility at the 2 m level becomes 400 m for V_{10} = 12, and 25 m for V_{10} = 22 m/sec. These values fit quite well with the estimates made by Loewe (1956) based on his observations and measurements made at PORT MARTIN in 1951. For instance, a "snow load" (DD) of 10 gram m^{-3} would correspond to V_{10} = 26 m/sec and a visibility of 10 m. Nevertheless, do not forget that the variability of the individual parameters from case to case can be quite large, and that the conditions along the coast of East Antarctica can differ substantially from those on the west antarctic plateau.

TABLE 3.30. Relationship between drift density DD and height above surface for two different values of wind speed at the 10 m level (V_{10}); rounded up from Budd et al.'s fig. 19, p. 95 (1966).

Height above surface	$V_{10} \doteq$ 12 m/sec	$V_{10} \doteq$ 22 m/sec
1 m	DD = 0.4 gram m^{-3}	8 gram m^{-3}
2	0.15 " "	3 " "
4	0.1 " "	1 " "
10	0.07 " "	0.7 " "

ii: Particle sizes. Drift particle replicas collected at BYRD by means of formvar-coated slides are shown by Budd et al. (1966, p. 82). Their evaluation indicated that the diameter of 85% of the particles gathered between 0.5 and 2 m above surface was in the size range from 60 to 120 μm, and slightly larger in the lowest half meter. (Only replicas > 20 × 20 μm^2 were collected.) Loewe (1956) gives 100 to 200 μm as most frequent size. In any case, that is considerably smaller than the majority of particles or flakes in an ordinary precipitation of snow. Occasionally, it has therefore been possible to confirm the contribution of real precipitation in blizzards. The small size of the drifting or blowing snow particles helps the strong winds to pack the snow deposits very tightly. At PORT MARTIN Loewe (1956) found that the surface deposits had a density of 0.4 to 0.5 (gram cm^{-3}), with no marked density increase down to a depth of 3.5 m.

iii: Drift snow transport. Across the storm-swept parts of the east antarctic coast, the amounts of snow transported by the wind can be very large. Loewe (l.c.) concluded that during the frequently recurring periods with blowing snow at least $50 \cdot 10^3$ kg of snow are carried from the continent across each meter of the coast line of the CAPE DENISON - PORT MARTIN region, per day. Budd et al. (ℓ.c) developed a formula for the drift snow transport under antarctic condi- tions as function of the wind, V_{10}, and applied it to the layer from surface to 300 m above it. Here are a few values, somewhat rounded up or off:

10 m level wind speed   10   15   20   25   30   m/sec
Drift wind transport (0-300 m)   10   30   80   230 $650 \cdot 10^3$ kg m^{-1} day^{-1}.

The numerical values in the formula are based, among other needed input, upon measurements of wind profiles in the lowest few hundred meters above BYRD Station. Fig. 3.10 (vertical wind profiles at MIRNY, MOLO, and NOVO) leaves no doubt that in many cases of strong katabatic flow at the east antarctic coastal stations the low level jet stream layer is very shallow; already at the 200 m level the wind speed is often much smaller than in the lower layers. That was not known when the Budd et al. (1966) formula was developed; it suggests that the above quoted values probably lead to an overestimate of the drift snow transport across the coastline. Naturally, the snow transport in the lowest 2 meter above surface can be determined more exactly. Fig. 3.26 shows the results of recent and earlier research.

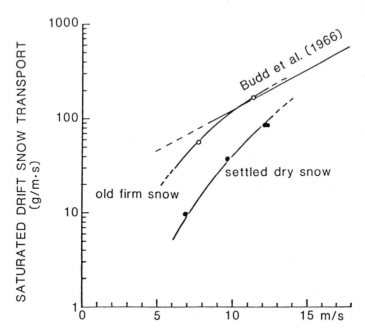

Fig. 3.26. Saturated drift-snow transport between the surface and the height of 2 m. The wind speed refers to 1 m height. From Takeuchi (1980) reproduced with the author's and editor's permission.

3.6 THE WINDCHILL PROBLEM

When it is cold and windy, it is easy to find out that neither temperature nor wind speed alone suffice to get a measure for the heat loss of the
human body in inclement weather. The cooling effect of a combination of the two
variables has been studied by Siple and Passel (1945) in a series of experiments at Little America in 1939-40, leading to the development of an empirical
formula for the relative cooling power of the air or the rate of heat loss (H)
of a person exposed to cold winds:

$$H = (10\sqrt{v} + 10.45 - v)(33 - T) \text{ kcal m}^{-2} \text{ hour}^{-1} \quad , \qquad (3.8)$$

with wind speed v in m/sec and temperature T in °C, 1 kcal m^{-2} hour^{-1}
≈ 1.16 watt m^{-2}. The 33°C is introduced as the neutral dry skin temperature.
With the assumption of a reference wind speed of 1.8 m/sec (4 mi/hour), a formula can be derived for the windchill temperature T_w (Falconer, 1968; Dare,
1981):

$$T_w = \frac{1}{22} (10\sqrt{v} + 10.45 - v) (T - 33) + 33°C \quad . \qquad (3.9)$$

Of course, the precision of the constant 10.45 is quite fictitious in view of
the possible exactness of the original measurements and the assumptions made.
The dependence of T_w on T and v can easily be presented in a nomogram as shown,
for instance, by Anthes et al. (1975). It is widely used in the U.S. by the
National Weather Service and TV meteorologists who on windy winter days in the
northern states enjoy the large negative values of T_w as a welcome and impressive warning to laymen. A few numerical values are given in Table 3.31.

For the experienced professional engaged in outdoor work in the
Antarctic, such a relative quantity like T_w is much more problematic. The main
difficulties, as discussed in detail by Court (1948), are: i. The formulas
cannot do full justice to the dry cooling process at high wind speeds; the term
$(10\sqrt{v} + 10 - v)$ reaches its maximum for v = 25 m/sec. ii. There are other processes contributing to the heat loss of the human body besides the windchill
effect described by equation (3.8); for instance, heat loss through the lungs
has been estimated to reach up to 20% of the total loss, and under specific
conditions also the radiative heat exchange can become important. That means
that windchill is not the only chill the human body experiences at temperatures
far below freezing. iii. a decisive point must be the protective clothing for
anybody working outdoors in the polar regions; that cannot be included in a
formula such as (3.8) or (3.9).

TABLE 3.31. Windchill temperature (T_w) as function of air temperature (T, °C) and wind (m/sec).

T		0	-20	-40	-60	- 80°C
Wind V = 5	T_w =	- 9	-34	-59	-84	-110°C
10		-15	-44	-73	-102	-131
15		-18	-49	-80	-111	-142
20		-20	-52	-84	-116	-147

In a discussion of the experience of polar explorers under extreme conditions of low temperatures and strong winds, Dalrymple and Frostman (1971) conclude that "high winds are probably a limiting factor (for outdoor activity) more than low temperatures", provided appropriate clothing is used.

Chapter 4

ATMOSPHERIC CIRCULATION AND ITS DISTURBANCES

Over the Antarctic continent and its ice shelves more than in other parts of the world, the coupling between the winds in the boundary layer and the flow of air above it is weak. This is due mainly to the boundary layer's pronounced static stability which makes possible the strong effect of the configuration of the terrain on the surface winds. Over the southern ocean, this effect disappears rapidly with increasing distance from the coast; then the connection between surface and upper winds cannot differ much from that over other oceans. Nevertheless, the cyclonic disturbances forming or intensifying in a strongly baroclinic field (i.e., a band-like area across which the isobaric temperature gradient in the troposphere is large), and the momentum transfer downward through an unstable boundary layer (cold air over relatively warm water), can produce over the southern ocean surface winds and a state of the sea surpassing what can be found on other oceans. This was well known hundred and more years ago when prior to the opening of the Panama Canal in 1914 the sailing ships went through the Drake Passage. In the book 'Master in Sail' (Learmont, 1950) one finds a quotation from an article by Mr. Huycke, as follows: "During the months of May, June and July of 1905 no fewer than 130 sailing-vessels left European ports for Pacific coast ports of North, Central and South America...Out of this number, 52 arrived at their destinations, four were wrecked, 22 put into ports in distress after Cape Horn damage, and 53 had not arrived or were unaccounted for by the end of November." In recent years, a dramatic description of the violence of the southern seas has been given by D. Lewis (1975) who sailed solo from Australia to the Antarctic Peninsula and on to Cape Town.

4.1 THE CIRCUMPOLAR VORTEX

The meridional temperature gradient between the coast and the open ocean near the surface is moderate in the summer and large during the rest of the year; it decreases with height (Table 4.1). More detailed data and charts of the mid-season average conditions have been given by van Loon (1964, 1966). In the monthly averages the sign of this gradient does not change all through the troposphere, though it can happen in the realm of individual strong cyclones and in the rare blocking situations (defined in section 4.2.6).

TABLE 4.1. Average meridional temperature gradient (°C/1000 km) northward from coastal stations at various levels of the troposphere in the mid-season months. Data for points 1000 km north of stations have been taken from Taljaard et al. (1969).

Month	Station	SAN	MOL	MIR	DUD	LAM	ELL/BEL
	Lat.	70	68	67	67	78	78°S
	Long.	2°W	45°E	93°E	140°E	163°W	40°W
JAN	Sfc	5	4	8	6	6	$4°/10^6$ m
	850 mb	4	4	6	6	4	4
	500	3	3	3	5	4	3
	300	3	2	1	3	1	2
APR	Sfc	18	10	18	14	21	15
	850	7	6	5	9	7	9
	500	4	4	3	6	6	5
	300	4	3	3	3	3	4
JUL	Sfc	18	11	19	16	22	15
	850	8	7	7	11	8	9
	500	5	5	4	6	5	4
	300	5	4	4	4	4	3
OCT	Sfc	7	10	17	14	14	10
	850	6	5	7	10	7	8
	500	4	4	5	7	4	5
	300	4	4	4	5	4	4

For any layer between two constant-pressure levels, a temperature[*] increase equatorwards means an increase of layer thickness in that direction, equivalent to a westerly (eastward directed) thermal wind. The latter is defined in Chapter 3. It follows that westerly winds increase with height, and easterly flow in a lower layer must under these conditions decrease with height and gradually change into a westerly one. It should be noted that in the free atmosphere the time-averaged geostrophic wind $\vec{V}_g$, in general differs very little from the resultant wind $\vec{V}$, while the same cannot be said of the instantaneous (or referring to a short time interval, of minutes or hours) vectors $\vec{V}_g$ and $\vec{V}$.

The geostrophic wind is directly proportional to the slope of the isobaric surfaces, divided by the Coriolis parameter which increases slightly with latitude. Based upon that simple relationship, the average flow pattern of the circumpolar vortex can be described easiest by the absolute topography of selected constant pressure surfaces, as shown in Fig. 4.1, maps c to f. The

[*]Note: To be exact, one must use the "virtual temperature", a variable that takes also the water vapor content of the air into account. However, in the range of temperatures here considered, the difference between the two variables is small enough to be disregarded.

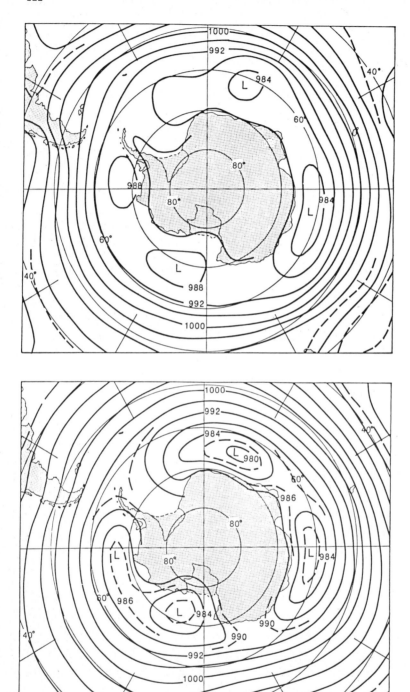

Figs. 4.1.a and b. Atmospheric pressure at sea level, averages for January (above) and July. Units: millibars; 1 mb = 100 Pascal.

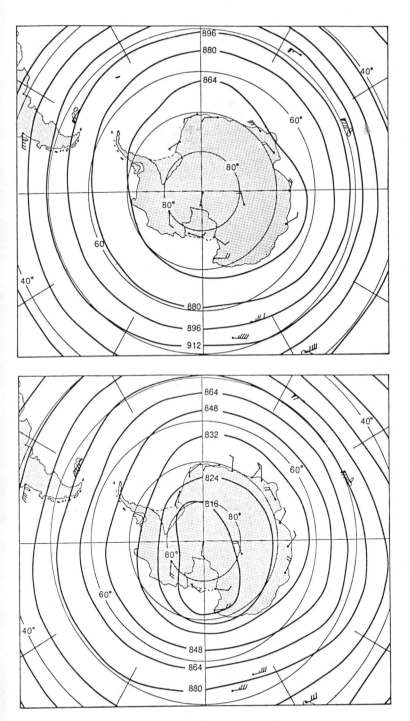

Figs. 4.1.c and d.   Average absolute topography (contour lines) of the 300 mb
surface, in January and July. Units: Geopotential decameter.

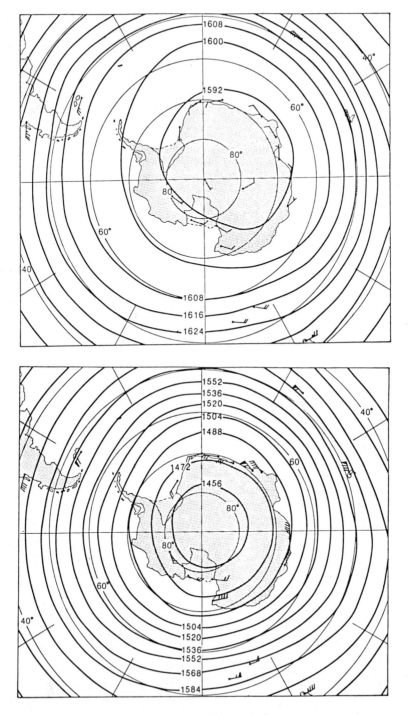

Figs. 4.1.e and f.   Average absolute topography (contour lines) of the 100 mb surface, in January and July. Units: Geopotential decameter.

direction of $\overline{V}_g$ is parallel to the contour lines. Only for the lowest level, i.e., mean sea level, the maps a and b show the pressure field itself, as generally done in synoptic meteorology and climatology in the past 150 years (Schneider-Carius, 1955, p. 248).

4.1.1 Position and vertical structure of the vortex over the continent

Except for a few mountain peaks in the Sentinel Range (pages 2-3), the first standard level over the continent above the atmospheric boundary layer is the 500 mb level. Near the vortex' center, the average height of that often used level is in summer about 150 m above 5000 m, in winter and spring 150 m below. The average position of the center is not sharply defined, though it can be taken for certain that it is not at 90°S; it can frequently be near the Ross Ice Shelf. Since there are too few sounding stations, an exact upper air analysis is not possible. Indirectly, the daily winds at the South Pole as shown in Fig. 4.2 can give a more realistic picture of the variability of the vortex' center. It must be understood that wind speeds greater than 10 m/sec at the Pole (observed on more than 50% of all days of July at 500 mb) are clearly incompatible with a polar-symmetric vortex. The latter is true also for the 300 mb level where at 10% of all days the wind speed exceeds 25 m/sec. In the winter 1972, the strongest tropospheric wind over the SOUTH POLE was 41 m/sec at 350 mb (3 August), the strongest wind shear in the 600 to 400 mb layer from 281° 18 m/sec to 208° 37 m/sec (26 July), corresponding to warm air advection on the order of 8°/3 hours.

There are good reasons to assume that the asymmetry of the atmospheric polar vortex is a consequence of the asymmetric topography of Antarctica. The resultant winds and thermal winds at two stations in the interior with adequate upper air records are shown in the graphs of Fig. 4.3 for the two winter months July and August combined. The average isobaric (or nearly horizontal) temperature gradients in the troposphere, orthogonal to the thermal wind, are clearly not in the direction of the meridians, and there is prevailing cold air advection over VOSTOK, warm air advection over BYRD. The latter is the case also for SOUTH POLE; it will be discussed in Section 5.3.2. Multi-annual average values of cold or warm advection for the two winter months are given in Table 4.2.

It must be noted that the heat budget of the free atmosphere for the polar night consists of only three major terms; translated into cooling or heating rates (degree/day) they represent the long wave radiation loss of heat (CL), advective heating or cooling (AD), and temperature increase or decrease by downward or upward vertical motion (VM). Since the average change of the real temperature from the first days of July to the last of August is negligible,

$$\overline{CL} + \overline{AD} + \overline{VM} \doteq 0. \tag{4.1}$$

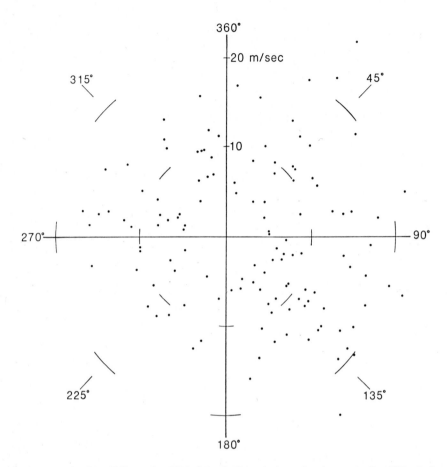

Fig. 4.2. SOUTH POLE, July 1971-75, daily winds > 5 m/sec at the 500 mb level. Altogether 126 soundings, corresponding to 87% of all soundings, 81% of all days. Imaginary straight lines from the coordinate center to each dot indicate magnitude and direction of the wind vector. A dot in the 135° (± 45°) quadrant stands for a day with the wind from the northwest quadrant, i.e., from the southern Weddell Sea region.

That means: over the interior of East Antarctica as represented by the station VOSTOK, where $\overline{CL}$ and $\overline{AD}$ are negative, the prevailing effect of vertical motion $(\overline{VM})$ of a stable air mass must be positive and relatively large. $\overline{VM} = +4°/day$ would approximately correspond to a sinking motion of 1 cm/sec. For West Antarctica as represented by BYRD station, in contrast, the average monthly or

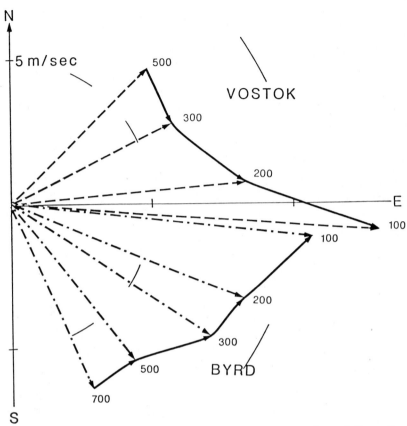

Fig. 4.3. Resultant winds (dashed and dash-dotted arrows) and thermal winds (solid arrows) in the free atmosphere up to the 100 mb level (about 14.5 km above sea level) over BYRD and VOSTOK, for July and August combined. Warmer air lies to the left of the thermal wind vector.

TABLE 4.2. Multi-annual average advective warming or cooling (°C/day) over West and East Antarctica in the troposphere above the boundary layer and lower stratosphere, July and August combined. p_S = average surface pressure; warming or cooling rate in °C per day.

Station ($\bar{p}_S$)	BYRD (802 mb)		S-POLE (677 mb)		VOSTOK (622 mb)
Layer	700 to 500 mb	1.6°/d	600 to 500 mb	1.7°/d	- - -
	500 to 300	1.5		0.7	- 1.1°/d
	300 to 200	1.6		0.3	- 2.2
	200 to 100	1.4		0.2	- 1.1
Years of data used	1957-68		1957-75, except 71		1958-60, 64-71

bimonthly value $\overline{VM}$ may not differ much from zero; the negative $\overline{CL}$ can be compensated by a positive $\overline{AD}$. This reasoning does <u>not</u> apply to individual weather situations. When there are well defined fronts, the value of AD can be a few degrees per <u>hour</u> at any place in the antarctic troposphere.

This explanation of a noticeable difference between the predominating budget terms $\overline{AD}$ and $\overline{VM}$ at East and West Antarctica can be supported by showing the mean lapse rates (vertical temperature gradients) of the 500 to 300 mb (5 to 8.5 km) layer for all available upper air stations (Fig. 4.4; also Kutzbach and Schwerdtfeger, 1967). The smaller lapse rate testifies for more frequent or more intense downward vertical motion. The pronounced sinking motion in the troposphere of East Antarctica can be attributed mainly to the boundary layer outflow from the interior to the adjacent ocean and ice shelves. A quantitative

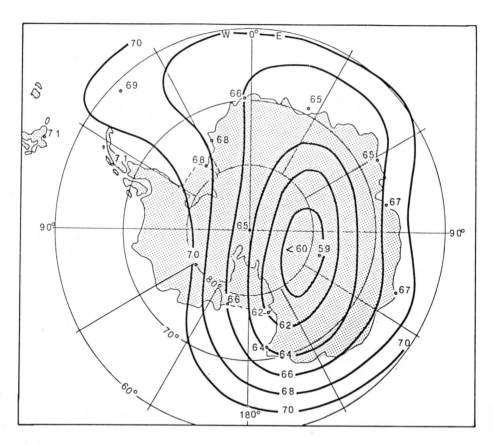

Fig. 4.4. Average lapse rate of the layer between 500 and 300 mb, for July and August, with ten or more years data since 1957 for most stations.
60 = 6°C/geopotential kilometer.

comparison of the outflow (per unit area) from East and West Antarctica cannot be given because not even one coastal station has been in operation (until 1980) between 80 and 160°W. Nevertheless, 12 years of observations at BYRD as well as snow accumulation data obtained by numerous traverses and glaciological studies (Vickers, 1966; Whillans, 1975) confirm that cyclonic disturbances approaching the continent from west-northwest with their prevailing upward vertical motion affect West Antarctica much more than the major eastern part of the continent.

4.1.2 Characteristics of the circumpolar vortex in subpolar latitudes

Monthly average maps like those in Fig. 4.1 can only give a general picture of the vortex. The most important questions left unanswered refer to the instantaneous fields of motion as shown in synoptic charts. For the middle and higher latitudes of the Southern Hemisphere, the lack of an adequate network of weather stations has always been the most serious impediment, though the International Geophysical Year 1957-58 (IGY) brought a decisive improvement, the meteorological satellite a few years later another one. Daily synoptic charts for the Southern Hemisphere, surface and 500 mb, have been elaborated for the IGY by the Weather Bureau of South Africa and published as Part III of the IGY World Weather Maps series. For a few subsequent years such maps can be found in the South African journal NOTOS. Later the International Antarctic Research Center at Melbourne, Australia, and then the National Meteorological Analysis Center of the Australian Bureau of Meteorology itelf took over. The Bureau's daily, in recent years twice daily, surface and 500 mb maps are available on microfilm; with the impressive development of satellite techniques in the 1970s, the completeness and continuity of these analyses have improved noticeably. Finally in 1979, "the buoys of FGGE" (explained on pages 9-10) brought the latest advance in meteorological technology. At present, many papers elaborating the new information appear in the meteorological journals; only a few can be discussed in Section 4.2.1.

Daily analyses of wind and temperature at the 200 mb level (about 11 km) have been produced, based on the results of the GHOST Project (page 10) combined with the regular radiosonde ascents made at land-stations. The Ghosts, free flying, constant density level balloons (superpressurized, constant volume) were launched by the U.S. National Center for Atmospheric Research (NCAR) from New Zealand (Solot, 1967; 1968). A constant density level near the 200 mb surface was chosen because at such elevation the lifetime of the balloons had proved to be greatest, on the order of a year. The trajectories of these balloons represent, with good approximation, the path of a mass of air in which the respective balloon is floating. To this purpose the choice of the 200 mb level was quite opportune since at that elevation the horizontal temperature gradients and vertical windshear are generally small, so that a slight sinking or rising motion of the air mass would be irrelevant for the determination of the wind. Fig. 4.5 shows the trajectory of a GHOST balloon in September and October 1966 which might be considered as typical. Other balloons have occasionally crossed and recrossed the equator (Solot 1967) or the antarctic continent (Solot and Angell, 1969), reaching a latitude of 85° (Schwerdtfeger, 1970a, fig. 21).

Another question not answered by the average charts for January and July in Fig. 4.1 regards the major seasonal variation of the strength of the westerly winds in the troposphere. The largest multi-annual monthly means appear near the time of the equinoxes, March-April and September-October (Schwerdtfeger and Prohaska 1955, 1956a, 1956b; Schwerdtfeger, 1970a; van Loon et al., 1971). This seasonal variation is clearly visible, without any statistical significance test, in Figs. 4.6 and 4.7 which show the monthly latitudinal averages at 55,

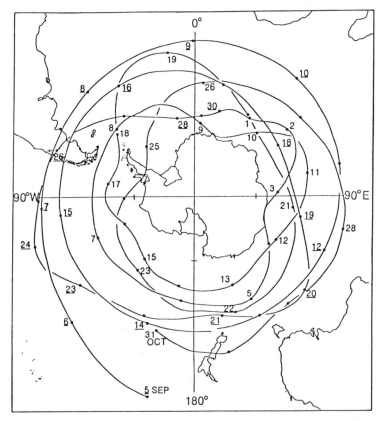

Fig. 4.5. Trajectory of a GHOST balloon floating near the 200 mb level (about 11 km) in September and October 1966. The numbers indicate the date, underlined for September (after Solot, 1967).

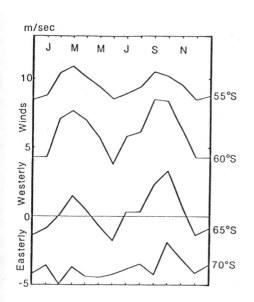

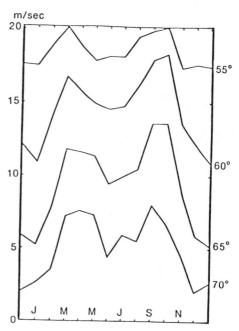

Fig. 4.6. Monthly latitudinal averages of the west to east (+) component of the geostrophic wind, computed from the sea level pressure field.

Fig. 4.7. Same as 4.6, computed from the 500 mb geopotential height values. Source of data for both graphs: Taljaard et al. (1969).

60, 65, and 70°S for the westerly component of the geostrophic wind in the lower and middle troposphere. The graphs confirm the adage of the equinoctial storms already known more than hundred years ago to the old sailors who navigated south of Cape Horn and Cape of Good Hope.

The phenomenon of the equinoctial wind maxima is related to the geometry of the Sun-Earth system, in the sense that the differential heating produced by the differing radiation budgets at the water surface between middle and polar latitudes is greater in spring and fall than in summer and winter (Schwerdtfeger, 1960). The resulting flow pattern in the lower troposphere is reflected in the pressure field: in middle latitudes equinoctial maxima, in polar latitudes solstitial maxima. More about the seasonal variations of pressure is said in section 6.4.

4.1.3 The stratospheric vortex

While the seasonal variation of the strength of the vortex in the troposphere tends to remain in modest limits, a different regime dominates in the
stratosphere. This change is quite visible in Fig. 4.8 which refers to an
individual point in subpolar latitudes. Above the 22 km level the westerlies
decrease from October to December by approximately 1 m/sec per day! With the
spring warming of the high layers (> 15 km) where the ozone content of the air
plays an important role, an anticyclonic system imposes itself for the two
months December and January. Table 4.3 shows how great this warming is in the
stratosphere near the Pole, in comparison to the troposphere over the continent
as well as the subpolar latitudes. The seasonal change of the vortex itself
becomes evident in Fig. 4.9, the meridional variation of the zonal winds at the
30 mb level in Fig. 4.10.

The question could arise why, repeatedly, reference is made precisely to
the 30 mb level. There is no meteorological reason to do so, rather the simple
fact that the number of radio soundings reaching still higher levels is small,

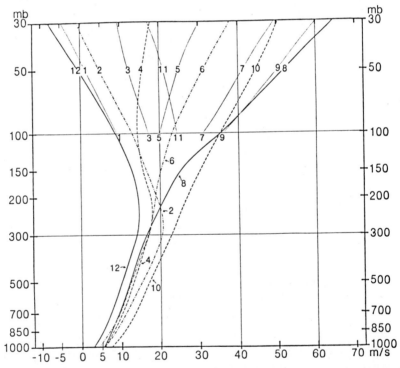

Fig. 4.8. Vertical profiles of the monthly mean west to east (+) geostrophic
wind component across the meridian 60°W, between 52 and 65°S, computed from the
height differences STANLEY - FARADAY. For the layer from 1000 to 100 mb, every
second month has been omitted.

TABLE 4.3. Winter and early summer average temperatures in stratosphere (30 mb) and troposphere (500 mb) at the S-POLE and 60°S (zonal means). Sources: ESSA and NOAA Publications (1962-77) Climatological Data for Antarctic Stations 1957-1975; Taljaard et al. (1969); Knittel (1976).

Lat.	Level	July and August		November and December		Diff.
90°S	30 mb	-91° at	20.9 gpkm	-33° at	23.8 gpkm	58°C
	500 mb	-45	4.87 gpkm	-38	5.04 gpkm	7
60°S	30 mb	-70	22.5 gpkm	-42	24.0 gpkm	28
	500 mb	-34	5.09 gpkm	-29	5.17 gpkm	5

particularly in winter. Furthermore, such upper air information including the winds, becomes less reliable with increasing height. Notwithstanding, it can be stated without reservation that the westerlies in the winter months, and the easterlies in December and January, increase with height beyond the 30 mb level.

Only a few upper air stations in the south-polar regions have produced also in the cold season series of sufficiently high-reaching soundings so that average 30 mb winds can be determined for time intervals of less than a month. Fig. 4.11 shows what is meant. The impressive decay by 70 m/sec from the 50th to the 67th pentad, and the slow regeneration of the cyclonic vortex appear clearly.

Equally remarkable is the meridional component v of the stratospheric winds. Even at high elevations the vortex is not circular, though it is very nearly pole-centered (Fig. 4.9), and neither is its temperature field. A summary of the seasonal means of v in the sectors with the strongest positive and negative values is given for 70° and 60°S, in the 30 mb and 100 mb level. A wave-number 3 pattern (three sectors with v > 0 and three < 0) which exists in the troposphere and lower stratosphere in middle latitudes (van Loon, 1972) appears to decrease poleward and with height. Table 4.4 makes it evident that there is a steady flow of cold, polar stratospheric air to lower latitudes in the Bellingshausen and Weddell Sea sector, and comparatively warmer air toward higher latitudes mainly over the eastern half of the east antarctic coast. The temperature difference between these two streams is greatest in winter and early spring.

It appears that the month of October has the strongest meridional wind components; this is also the month in which the pronounced increase of mass weighing upon the continent (Fig. 6.8) begins. Therefore, this month has been chosen to show the temperature T at the 30 mb level as function of longitude, in Fig. 4.12. The temperature difference between sectors at 70 and 60°S is quite pronounced, as large as 20°C. This certainly confirms the (qualitative)

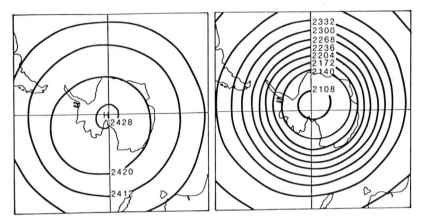

Fig. 4.9. Average height (geopotential decameter) of the 30 mb surface in January 1969-1973, and July 1968-1972 (after Knittel, 1976).

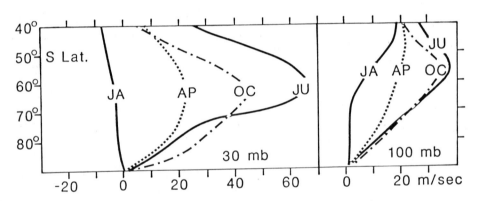

Fig. 4.10. Meridional profiles of the zonal average west to east (+) wind component at the 30 mb and 100 mb levels in the four mid-season months, from data for April 1968 to March 1973. Modified version of graphs published by Knittel (1976).

validity of the geostrophic wind data derived by Knittel (1976) for grid points south of 30°S. Still, there can be some skepticism, in particular for the western half of the coastal band of Antarctica where Knittel's values are probably too large. In contrast, on the east side from NOVOLAZ (12°E) to LENINGRADSKA (159°E) there are seven well distributed stations whose records make values for the 30 mb level in Table 4.4 relatively reliable. The results of rawinsonde wind measurements at several of these stations in the late 1970s are in agreement with Knittel's values published in 1976.

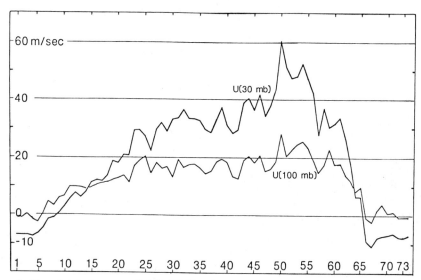

Fig. 4.11. Pentad means (1973-1977) of the west to east component (+) of the wind at the 100 mb (between 14.8 and 16.0 km) and 30 mb (21.5-24.2 km) levels over SYOWA STATION, 69°S, 40°E. Source of data: JARE REPORTS.

TABLE 4.4. Seasonal average meridional wind component $\overline{v}$ (m/sec) in sectors of 60° longitudinal width where $|v|$ is maximum, at 70 and 60°S, 30 and 100 mb level. Source: Knittel 1976.

	DEC - FEB		MAR - MAY		JUN - AUG		SEP - NOV	
Lat.	$\overline{v}$	Long.	$\overline{v}$	Long.	$\overline{v}$	Long.	$\overline{v}$	Long.
70°S	-1.8	(90-150°E)	-3.0	(90-150°E)	-5.0	(90-150°E)	-9.2	(70-130°E)
30 mb	+1.9	(90- 30°W)	+5.6	(110-50°W)	+8.5	(90- 30°W)	+9.7	(110-50°W)
60°S	-1.2	(90-150°E)	-2.2	(90-150°E)	-4.4	(90-150°E)	-7.4	(70-130°E)
	+1.5	(110-50°W)	+4.3	(110-50°W)	+5.8	(90- 30°W)	+7.9	(130-70°W)
70°S	-1.8	(110-170°E)	-2.7	(130-190°E)	-3.0	(110-170°E)	-4.9	(90-150°E)
100 mb	+2.7	(90- 30°W)	+3.3	(90- 30°W)	+4.0	(70 - 10°W)	+3.9	(90 -30°W)
60°S	-2.3	(90-150°E)	-2.3	(110-170°E)	-3.7	(90- 150°E)	-4.0	(90-150°E)
	+2.7	(90- 30°W)	+2.4	(90 - 30°W)	+2.8	(70 - 10°W)	+3.1	(90 -30°W)

136

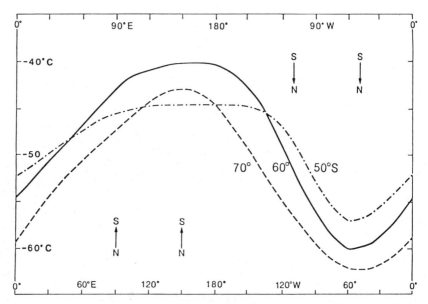

Fig. 4.12. Mean temperature of October at the 30 mb level (about 23 km) at 50°, 60°, and 70°S latitude as function of longitude. The arrows shall only indicate in which sector the positive (S→N) and the negative (N→S) v components are strongest; seasonal numerical values are given in Table 4.4. Source of data: Knittel (1976).

"With a low and falling glass
soundly sleeps the careless ass."
(old sailor's rhyme)

4.2 CYCLONES AND ANTICYCLONES OVER THE SOUTHERN OCEAN AND THE COASTAL REGIONS OF ANTARCTICA.

A description of the circumpolar vortex as given in the preceding Section would be inadequate or even misleading if it were not accompanied by a discussion of the individual, from day to day varying, disturbances: the cyclonic and anticyclonic systems. These can best be illustrated by a comparison of a mosaic satellite picture and an approximately simultaneous 500 mb analysis with the multiannual average 500 mb chart for the same month (Fig. 4.13a, b, and c). The date of a and b has not been chosen with the intention to present any extraordinary cloud vortex formations, but rather for the completeness and quality of the mosaic, that means "at random" from the meteorological point of view. Statistics of the frequency of cyclones over the southern ocean and the coastal regions of the continent will not be included in the following sections. It appears (to the author) that mostly due to the remarkable augmentation of satellite information and improvement of the respective evaluation techniques, a

Fig. 4.13a. Mosaic of cloud pictures in the infrared (daytime), from satellite NOAA 4, orbits 3913 - 3926, 24 - 25 September 1975.

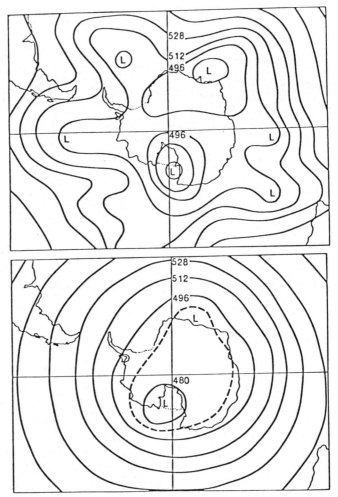

Fig. 4.13b. 500 mb contour lines (geopot. decameters) for 00z 25 September 1975, slightly modified version of the Southern Hemisphere analysis of the Australian Bureau of Meteorology.

Fig. 4.13c. Multiannual average 500 mb contour lines for the month of September, modified from the chart of Taljaard et al. (1969).

homogeneous series of baric synoptic charts just does not exist. Additionally, it must be considered that the year to year variations of cyclonic activity, for instance the second against the first year of the IGY, or 1979 against 1978, can be very large, so that averages and frequency statistics for only a few years can hardly claim much significance.

4.2.1 Cyclones, limitations of analysis

There can be no doubt that in the southern hemisphere the maximum cyclonic activity has to be expected in the circumpolar band of low pressure that so clearly (Fig. 4.1a and b) exists in the seasonal and annual _average_ sea level pressure maps. This nearly zonal band is the locus of the highest frequency of cyclones or their greatest intensity, or both. The cyclones can be of all possible types, fast migrating to stationary, shallow with relatively warm air in their central part, or cold lows intensifying with height through the troposphere.

All these transient systems have a very important function: the meridional exchange of different air masses, continental cold air moving toward lower latitudes, warmer and moister air toward the continent. In fact, the transport of maritime air poleward is the main mechanism of moisture supply, without which Antarctica would have an entirely different climate and topography. There is more about the provision of moisture in chapter 5. With regard to the temperature regime one can argue that a decrease of the activity of the subpolar cyclones will contribute mostly to lower temperatures on the continent. This, however, can be only the first step in a chain of events. Since the temperatures over the ice-free parts of the southern ocean will change much less than over snow, ice, or land, there will be an increase of the meridional temperature gradient. This again will help to re-intensify the cyclonic activity, so that Nature keeps the range of variations within certain limits, by means of a relatively simple, negative feedback arrangement.

The average latitude of the low pressure belt surrounding the continent is approximately 65°S, probably somewhat lower (62°S) north of the coast of East-Antarctica between 30° and 150°E longitude. There are also seasonal variations, provable at least in the Weddell Sea sector where several stations with long records exist in the critical latitudes, 50 to 65°S.

Coast-parallel migrating cyclones, with centers a few hundred kilometer off shore, are the most frequent; in fact, they can be seen practically every day on the synoptic weather charts. Naturally, the passage of these cyclones is combined with changes of weather conditions along the coast. The effect on the strength of katabatic winds has been discussed in Section 3.2. In parts of the coast favorable for this kind of surface winds, it is entirely possible that the maritime air of the warm sector of the cyclone does not displace the cold continental air near the surface. Nevertheless, above the boundary layer the relatively warm and moist air can cross the shore line and produce intense upslope precipitation.

Among the cyclonic disturbances which appear in the subpolar latitudes, there is a fair number of well-defined low pressure systems originating in the lower middle latitudes over the warm ocean currents east of the major land masses of the southern hemisphere, and moving east-southeastward. That is: from the region north of New Zealand east-southeastward to the Drake Passage or Bellingshausen Sea, others from the southwestern Indian Ocean into the northern Ross Sea area and from the western South Atlantic off the coast of southern Brazil to the eastern end of the Weddell Gyre north of Enderby Land (Streten and Troup, 1973). It appears that many of these systems intensify south of 50 (± 5)°S latitude. Normally, such systems can generate secondary cyclones or so-called families of cyclones.

One of the early investigations of cyclones over the southern ocean by means of satellite pictures in the visible and infrared was carried out by D.W. Martin (1968a and b). Some of his conclusions, confirmed by later studies based on more advanced satellite technology, may here be mentioned, partly verbatim.
i: Among southern ocean cyclones there is great variety in genesis, movement, size and intensity; consequently, there are frequent deviations from the classical wave cyclone model. For instance, frontogenesis appears often to occur with cyclogenesis, not prior to.
ii: As a cyclone intensifies, its vortex cloud roughly doubles in size each day. Growth ceases as a cyclone begins to weaken. Decay of a cyclone is indicated by increasing distortion of the visible vortex, and by a lowering of cloud tops. In an advanced stage of decay over the open sea the vortex can degenerate into a field of intense low-level convection (cold air over warm water).
iii: As everywhere, vertical motion and horizontal advection are the most effective processes changing a given cloud field in the free atmosphere. Therefore, the difference between the actual change (seen from satellite pictures) and the advective change (if obtainable from the horizontal wind field) can be attributed to vertical motion.
iv: There is no convincing evidence to show that southern ocean cyclones differ in any fundamental sense from their counterparts over the northern waters.

There is one aspect of cyclogenesis or deepening, and cyclolysis or filling, which is often not sufficiently taken into account: The atmospheric pressure measured at a place of constant height can decrease with time only if the mass of air in the column above the place decreases. When there is a cyclonic flow pattern near the surface, friction induces convergence of mass in the boundary layer, a process that can lead to upward vertical motion and initially also to an increase of pressure. If, notwithstanding, the pressure at surface is decreasing, there must be a greater loss of mass somewhere higher up in the atmosphere, by means of divergent flow, or ageostrophic advection, or both. This is one of the reasons why detailed information on the state of the upper air is so important for synoptic meteorology. It also explains why in the Antarctic a detailed, reliable analysis of formation or intensification of cyclones cannot be carried out. With a distance of a thousand or more kilometers from one upper air station to the next, fields of divergence and vorticity in various layers of the free atmosphere would remain manifestations of the analysts' or programmers' guess, not necessarily what the dynamics and thermodynamics of the real conditions require.

Since not all nascent cyclones become visible on satellite pictures in an early state of development, it is often impossible to exactly determine the area in which a "cyclogenesis" has taken place. Nevertheless, it seems certain that cyclones of polar origin, generally of smaller size than the mid-latitude immigrants, form over the coastal waters or ice fields where the contrasts between relatively warm, moist maritime air and much colder continental air are large. In the meteorological literature there are various studies regarding cyclogenesis, frequency of centers per area, cyclone tracks, speed of motion of the systems, etc. For the latter, 15 m/sec appears to be an acceptable median value for the belt between 40° and 60°S. An extensive, easily accessible review of these topics has been given by J.J. Taljaard (1972) in the Am. Met. Soc. Monograph (35) on the Meteorology of the Southern Hemisphere. Taljaard points out that cyclogenesis frequently occurs in high latitudes, near the coast of the Indian and Pacific Oceans. The results of case studies of the development of cold air mass cyclones (Mullen, 1979), and cyclogenesis in polar air streams (Reed, 1979), both authors referring to north-polar regions, support Taljaard's concepts. In this context, also a theoretical study (Mechoso, 1980) that includes consideration of the antarctic topography, is of interest. With some simplifying but not unrealistic assumptions Mechoso shows that the combination of baroclinic waves approaching the coastal regions from middle latitudes with the effects of the topography of the continent and the strong meridional temperature gradients can lead to the generation of westerly winds with a jetlike structure in the upper half of the troposphere over the coast or the steep slopes nearby. He concludes that in the region around Antarctica favorable conditions exist for local cyclogenesis. This is a relatively new perception, and it is not yet known whether it is confirmed by the complete evaluation of the FGGE data of 1979. Much new information and understanding of the cyclones over the southern ocean and the antarctic coast will become available within a few years. Therefore, in the following Sections, more attention will be given to the observed facts and the product, the weather, rather than to the producer, the cyclone.

Only as an example of existing uncertainties, Fig. 4.14 shows the sea level pressure map for 00z 11 July 1979, analysed at the National Meteorological

Analysis Center of the Australian Bureau of Meteorology (Guymer and LeMarshall, 1980, 1981). There appear five low pressure centers in the circumpolar belt, four of them lower than 950 mb, the deepest (at 69°S, 10°W) < 932 mb. This map is the product of manual analysis methods used by experts in southern hemisphere synoptic meteorology. It is based on the usual synoptic weather reports from all available land-stations and a few ships, _plus_ about 250 reports from the drifting buoys. For the same date and hour, and presumably with the same basic information, another sea level pressure map has been elaborated by "objective", i.e. numerical, methods at the European Center for Medium Range Weather Forecasts (FGGE, 1981). In that analysis, otherwise in fair agreement with Fig. 4.14, the cyclone at 69°S, 10°W reaches 950 mb, at best. It might be assumed that such discrepancies can soon be eliminated.

The average depth of the entire circumpolar trough on 00 11 July 1979 is 963 mb in the Australian analysis, and 964 mb in the European. That may be compared with the multiannual average trough-depth for the month of July based on all data available prior to 1967 (Fig. 4.1b), amounting to 984 mb, a value that is confirmed by a more recent average July map 1973-1977 (LeMarshall and Kelly, 1981).

Fig. 4.14. Sea level pressure chart, 00z 11 July 1979, as analyzed at the Nat. Met. Analysis Center, Australian Bureau of Meteorology, published by Guymer and LeMarshall (1980). The isobars over the antarctic plateau can be disregarded (see Section 4.3.1). Courtesy of the Director, B. of M., Melbourne.

A difference of 20 mb between a monthly mean and an individual day as average all around the continent is certainly an exceptional case, and it appears to "point to a need for reappraisal of the pre-FGGE surface pressure climatology" (Guymer and LeMarshall, 1980, 1981). That is probably justified, though it must be considered, as Trenberth and van Loon (1981) show conclusively, that the atmospheric circulation in the extratropical southern hemisphere was anomalous during most of the FGGE, December 1978 - November 1979. Particularly in July (and September) 1979, the westerlies between 40° and 60°S were exceptionally strong (Streten and Pike, 1980) in comparison with previous years, and the pressure at several East-Antarctic coastal stations was far below normal; even on the Antarctic Plateau July 1979 brought a record low monthly average (Table 4.5). One can conclude that the very low pressure values in the

TABLE 4.5. Multiannual and extreme monthly average pressure ($\bar{p}$, mb) at station level on the high plateau, and at sea level at two coastal stations, for the month of July.

Station	SOUTH POLE	VOSTOK	HALLEY	CASEY
Period	1957 - 80	1958 - 80	1956 - 82	1957 - 82
Average $\bar{p}$	677	621	990	985 mb
Standard dev.	5.5	5.0	5.6	6.1
Max $\bar{p}$	692 (1964)	631 (1964)	1008 (1964)	1001 (1964)
Min $\bar{p}$	665 (1979)	609 (1979)	979 (1979)	972 (1979)

center of four cyclones on 11 July 1979 are compatible with the average July pressure map (Fig. 4.1b) drawn 15 years ago by Taljaard et al. (1969). The FGGE results just seem to give a good example of extreme, not average, conditions.

The values mentioned, or listed in Table 4.5, already suggest that the interannual variations can be large, or simply, that the pressure field in one month can differ considerably from that of a year before, in spite of the fact that a circumpolar trough with prevailing westerlies north of it, and easterlies between the trough line and the coast, practically always exists. Still, time series of less than 30 values cannot produce very convincing results. Therefore, more complete pressure statistics are given in Appendix A3 for the only station which has a continuous record since the early years of this century and is located near the band of maximum cyclonic activity.

In this context it is also of interest to see what are the absolute maximum and minimum sea level pressure values that have been measured at various places along Antarctica's coasts within a period of years. Unfortunately, such values are seldom published, and only short series of absolute extreme values of pressure have become available for most stations. Therefore, the extremes given in Table 4.6a are for the four seasons instead of each month. For MAWSON (1957-1968) a different kind of statistics is available in the form of a frequency count of all pressure values below certain limits; it is shown in Table 4.6b.

TABLE 4.6a. Absolute maximum and minimum pressure (at sea level) of various coastal stations. Because of the shortness of the available series of data, the extreme values are given only for the four seasons (DJF = December, January, February).

Station	Lat.	Years		DJF	MAM	JJA	SON
LIT. AMERICA	78.6°S	6	max.	1020	1027	1031	1023 mb
1929-58			min.	969	952	935	932
			diff.	51	75	96	91
SCOTT	77.9	16		1017	1035	1034	1023
1957-72				964	955	950	945
				53	80	84	78
SHA and ELL	77.8	7		1019	1031	1029	1019
1956-62				968	954	945	946
				51	77	84	73
HALLEY	75.5	15		1014	1024	1028	1015
1956-70				955	951	951	939
				59	73	77	76
HALLETT	72.3	7		1017	1027	1029	1017
1957-63				965	961	952	950
				52	66	77	67
BAUDOUIN	70.4	6		1010	1022	1024	1023
1958-66				955	946	950	945
				55	76	74	88
SANAE	70.3	11		1014	1019	1025	1021
1960-70				964	945	941	945
				50	74	84	76
ADELAIDE I.	67.8	11		1017	1026	1027	1019
1960-70				956	944	941	937
				61	82	86	82
MOLODEZ	67.7	11		1011	1017	1023	1022
1963-73				953	956	947	941
				58	61	76	82
MIRNY	66.6	18		1015	1022	1026	1016
1956-73				952	945	939	942
				63	77	87	74
WILKES	66.3	7		1025	1016	1026	1015
1957-63				952	940	949	947
				73	76	77	68
FARADAY	65.3	24		1018	1033	1035	1026
1947-70				953	936	937	935
				65	97	98	91
ESP and HOPE	63.4	12		1018	1024	1032	1024
1945-59				957	958	943	947
				61	66	89	77
ADM. BAY	62.1	13		1017	1030	1034	1025
1948-60				959	958	939	946
				58	72	95	79

TABLE 4.6b. Numbers of three-hourly observations with the sea level pressure less than 951 mb, derived from the published station level data. MAWSON, 1957-1968.

Total number of all observations	Season 35,064	< 951 mb	DJF 2	MAM 6	JJA 5	SON 22

The minimum values listed in Table 4.6a are by far not the lowest that can occur. FOSSIL BLUFF Station was in operation many summers, but only one full year, 1968; in that September, the sea level pressure decreased twice to values less than 940 mb. Much more impressive, however, is the pressure observed and recorded at PORT MARTIN, 3 September 1951, reproduced in Fig. 4.15: P_0 = 926.9 mb, with T = - 2°, V = 14 m/sec from the east (Prudhomme and Le Quinio, 1954; Le Quinio, 1956). This appears to be, up to now, the lowest pressure value reliably determined on any part of Antarctica's coast. Second in line seems to be, with 930 mb, ORCADAS in April 1912. It is most likely that on the southern ocean itself occasionally lower values occur. The FGGE data of the drifting buoys strongly suggest it.

It might also be remarkable that the extreme maximum pressure of PORT MARTIN was measured 16 June 1951: P_0 = 1029.7 mb, T = - 17°, V = 2 m/sec from SSW. It so happens that PORT MARTIN is the only coastal station with the total range of measured P_0 values greater than 100 mb. ORCADAS Island is the only other station for which such a range has become known (Table A3, last page).

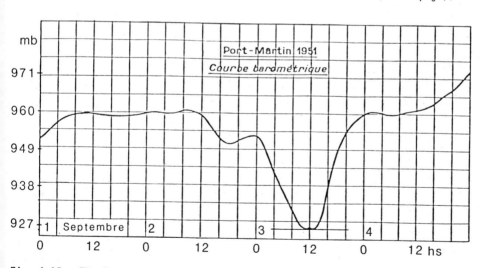

Fig. 4.15. The lowest pressure (at sea level) ever recorded at an Antarctic station.

4.2.2 About the vertical extent of some cyclonic disturbances
in coastal regions

It is often found that in the realm of an intense mid- or higher latitude low pressure system the strongest winds occur in the upper troposphere,
generally on the equatorward side of a mature cyclone's track. Still, it might
be worthwhile to point out where, and under what conditions, the strongest
winds exist in the lower troposphere. When a deep depression, moving or not,
remains north of the coast of East Antarctica, as it happens quite frequently,
the easterly winds between the cyclone's central region and the coast increase
with height through the boundary-layer as the friction effect diminishes, but
not much farther. Above that layer, the air of the cyclonic vortex over the
ocean can be much warmer than the air to the south which either was already
previously located over the edge of the cold continent or is air of maritime
origin which recently has been carried southward and uphill, cooling by lifting
and radiative heat loss. Under such conditions there must be a coast-parallel
thermal wind with a strong west to east component between the central part of
the cyclone and the inland in the layer 1000 to 3000 m above sea level, approximately; that means that the easterly component of the real wind has to
decrease with height and eventually change into a westerly.

A development that fits into this picture is shown in Fig. 4.16. In the
two days with lowest pressure and strongest northeasterlies, the speed maximum
was near 850 mb (about 950 m), with 33 m/sec from NE as average for 4 soundings
17 and 18 July. At 500 mb (4750 m) for the same time the speed was 14 m/sec,
at 300 mb (8050 m) 11 m/sec; finally, at 100 mb (14,600 m) there were only
westerlies, well characteristic for the winter. It is noteworthy that in the
lowest 2000 m the wind direction changes very little though the sea level pressure decreases by 39 mb (day 15 to 17), then rises by 43 mb (17 to 19). Apparently a strong, but shallow, cyclone became stationary to the northwest of
Lützow Holm Bay (69°S, 38°E) and decayed fast. The DMSP-S infrared satellite
maps for this region, 16-18 July support such interpretation. The development
here described is not a rare event. Fall (and winter in regions with a small
width of the packice belt) probably are the preferred seasons, because then the
temperature contrast between ocean and continent can be sharpest.

4.2.3 Cyclonic disturbances extending across mountains

When an intense cyclone moves east-southeastward across the Bellingshausen Sea and approaches the mountains of the Antarctic Peninsula south of 70°
latitude (with elevations between 2000 and 3000 m), it will not be able to continue its march undisturbed. To show what might happen, a case of the year 1968
has been selected because for this year data of FOSSIL BLUFF Station are

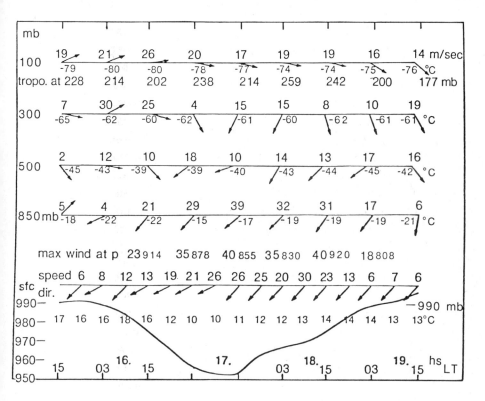

Fig. 4.16. Surface and upper wind, and temperature, at SYOWA Station (69°S, 40°E), during approach and decay of a strong depression which exists mainly in the lower troposphere, northwest of Lützow Holm Bay, 15-19 July 1974. In the "max wind at p" line the large number is the wind speed, the small number is the pressure level at which the maximum was measured.

available. With the synoptic observations of three stations along the west coast of the Antarctic Peninsula, FAR at 65.3°S, ADE at 67.8°S, and FOB at 71.3°S, it is easy to deduce at what latitude the cyclone's center comes to lie when near the coast, and to observe the decay or bipartition of the cyclone (Schwerdtfeger and Amaturo, 1979).

Special attention will be given to two aspects of these, sometimes quite formidable (center pressure < 940 mb), storms: i: The effect on the sea ice on both sides of the peninsula, and ii: The alteration of the cyclone itself by the transport of relatively warm and moist air masses toward the southwest corner of the Weddell Sea and farther south to the steeply rising slopes of the continent. Both effects can best be explained on hand of the example of 12 to 14 September (Figs. 4.17a and b).

i: As long as a sizeable positive pressure difference ADE-FOB persists, the vector representing the stress exerted by the surface winds on the ice must

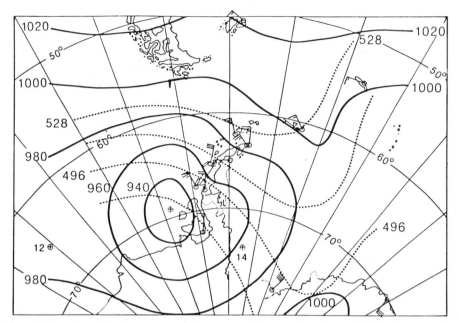

Fig. 4.17a. Example of a deep cyclone over the southeastern Bellingshausen Sea, 00z 13 September 1968; central pressure about 935 mb. The small circles at 65, 69, and 74°S indicate the estimated location of the center at 00z, September 12, 13, and 14. Dotted lines: Contourlines of the 500 mb surface, in geopot. decameters.

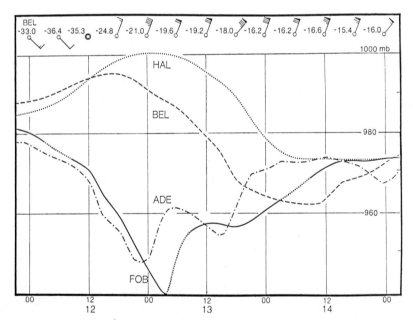

Fig. 4.17b. Barograms of two stations (ADELAIDE ISLAND and FOSSIL BLUFF) on the west side and two (HALLEY and BELGRANO) east of the Antarctic Peninsula, 00z 12 to 00z 15 September 1968. In the upper part: wind and temperature at BELGRANO.

have a large west-to-east component. A difference of 10 mb corresponds to a geostrophic wind of 15 m/sec (assuming air density 1.20 kg m^{-3} and mean latitude 70°) across the 400 km line between the two stations. On the Bellingshausen side, the stress is directed on-shore, essentially into Marguerite Bay (see Fig. 4.17a) where the ice conditions[*] must deteriorate depending, of course, also on the duration of the wind storm from the west quadrant; in 1968 the longest period of uninterrupted westerlies between ADE and FOB was eleven days, 23 September - 3 October. The preferred part of the year for this phenomenon must be the spring months, because that is the time when the mid-latitude high pressure ridge over the South Atlantic is strongest and extends farthest south (Table 4.7).

The wind stress on the Larsen Ice Shelf, under conditions such as shown in Fig. 4.17a, acts in the off shore direction, and is capable of producing leads and polynyas along the rim of the shelf. In recent years the latter phenomenon has been well documented by satellite pictures (Sissala et al., 1972; Gloersen, 1973; Kyle and Schwerdtfeger, 1974; Colvill, 1977). Indeed, this kind of synoptic situation appears to be the only one that can initiate major temporary changes in the ice cover of the western Weddell Sea (Section 3.4.2).

TABLE 4.7. Relative frequency (%) of different classes of the sea level pressure difference between ADELAIDE ISLAND and FOSSIL BLUFF (distance 400 km), in four parts of the year 1968. Observations for 00 and 12z; data for the last eleven days of December missing. A value of 10 mb corresponds to a geostrophic wind component of 15 m/sec; assumed density 1.2 kg m^{-3}.

	n	<-12.0	-11.9 to -8.0	-7.9 to -4.0	-3.9 to +3.9	4.0 to 7.9	8.0 to 11.9	⩾12.0	mb
Nov. to Jan.	162	-	2	8	80	9	<1	-	%
Feb. to May	242	-	1	17	76	5	<1	-	%
June to Aug.	184	<1	2	12	64	16	5	<1	%
Sep. to Oct.	122	<1	1	10	52	26	7	4	%
Jan. 1 to Dec. 20	710	0.3	1.4	12.3	70.2	12.4	2.6	0.8	%

zonal geostrophic   from E ⟵————⎦   ⎣————⟶ from W
wind component
Extreme values June 26: - 19.5 mb ≈ 28 m/sec
             Sep. 28: + 16.9 mb ≈ 24 m/sec

---

[*] regarding the ice conditions in Marguerite Bay prior to 1962 see Heap (1964), for the years 1951-1961 a detailed study by De La Canal (1963), and for the favorable years 1970-75 Schwerdtfeger (1976).

ii: An alteration of the cyclone itself is initiated when a southeastward progressing intense cyclone center approaches the southeast corner of the Bellingshausen Sea and the cyclonic pressure field reaches far eastward. It follows that there is a broad stream of relatively warm and moist air masses in the lowest few thousand meters of the atmosphere. This stream, clearly visible over the central and southern Weddell Sea, is heading toward the steeply rising slopes of the southernmost part of the peninsula. Since these air masses come to move over colder and colder surfaces, their vertical structure must be essentially stable. When they approach the continental escarpment, there will be "damming up" and a turning of surface winds to the right, resulting in convergence and positive vertical motion, all this in a form quite similar to that described in earlier sections for the case of easterly flow toward the mountainous peninsula, but here with more complications because of the greater irregularity of the terrain. Consequently, along and on the slopes one must expect not only considerable snowfall but, more important, an increase of surface pressure. This pressure effect, already mentioned in Section 3.3, can explain the observation that deep depressions which come into the southeastern Bellingshausen Sea and seem to advance southeastward, have much higher central pressure values when they reappear in the sea level pressure maps over the southern Weddell Sea or the Ronne Ice Shelf.

Altogether, strong meridional pressure differences between 67° and 71°S near 70°W have a remarkable impact on the ice conditions and the weather development at both sides of the Antarctic Peninsula. The annual snow accumulation in these regions is probably larger than indicated in the most valuable publication of Bull (1971) and others. That remains, however, no more than a conjecture as long as so little solid meteorological information is available. One sees immediately how useful a few automatic weather stations could be.

Modification of an intense cyclone and retardation of its advance is not restricted to the southern Antarctic Peninsula region. It can be observed wherever the configuration of the coastline and the topography of the hinterland are appropriate. Such disturbances tend to maintain, or reproduce, strong on shore wind components, warm air advection, and copious precipitation. An impressive storm of this kind in 1957, in the southeast corner of the Ross Sea near LITTLE AM. V, was experienced and described by Alvarez and Lieske (1960).

4.2.4 Early autumn storms in the northwestern Weddell Sea

The extreme shortness of the summer on the high plateau has been discussed in Section 2.3. Here it must be added that in the southern and western Weddell Sea, more than in other parts of the low level Antarctic, the relatively tempered summer can be, and in many years is, equally short. Two famous ships that later became victims of the Weddell Sea were beset by the ice very early: Nordenskjöld's Antarctic on 12 February 1903, Shackleton's Endurance on 19 January 1915. Only the nine months captivity of Filchner's Deutschland began relatively late, 6 March 1912. Since 1955, several of the annual passages of the Argentine icebreaker General San Martin to the base BELGRANO have confirmed that already in the second half of January the temperature- and wind-determined difficulties with the ice in the southern Weddell Sea increase rapidly.

Apparently, the topography facilitates the flow of cold air from the interior of the continent to the Filchner and Ronne Ice Shelves and so to the Weddell Sea ice. Then the barrier wind effect along the east coast of the Antarctic Peninsula helps bringing the cold air with strong southerly winds to the northwestern part of the Weddell Sea, at a time when on the other side of the peninsula the ice conditions, if any, are the most favorable of the entire year.

In February 1975, at the remarkably low latitude of 64°S, the early winter storms tried again to find a victim, this time without success. The endangered ship was a sturdy icebreaker, the above mentioned General San Martin. What happened is best described in the words of an officer-meteorologist of the Armada Argentina who was on board:

We left Marambio on February 21. At that time the sea was covered by 9/10 of sea ice and growlers, with several tabular icebergs in between. The navigation was extremely difficult. The ship could barely make any progress; after 48 hours the distance from the island (Marambio = Seymour Island) was not more than 5 to 6 miles. Then the number of bergy bits and medium size floes increased, and several strongly eroded icebergs appeared.

The 26th, our location being 15 to 20 miles north of Marambio, a windstorm from SW came up, the windspeed oscillating between 40 and 50 knots, ice continuously striking against the ship. It was an impressive spectacle, the ice floes and bergy bits piling up and rising one over another, forming mountains of ice. The pressure exerted by the ice broke some reenforcements of the ship's body and in some moments lifted the entire ship. Under these conditions any maneuvering was impossible, but then the commander took advantage of an aperture between two bergs to find some protection on the east side of one of them. This diminished the danger presented most of all by the smaller, eroded icebergs which had a height of 20 to 50 m and a diameter of 200 to 300 m and more, that is, a multiple of the size of the ship. The storm persisted for 48 hours. When the wind abated, the sea surface had changed into a compact mass of ice from which the icebergs and large pressure lines protruded. It resembled a lunar landscape.

For one week, the ship remained immobile. Reconnaissance by helicopter indicated that slowly some flaws and kind of channel were forming. The 2nd of March one could observe that a channel which seemed to be connected with other flaws had opened at stern. With the help of explosives and several other maneuvers it finally, after one week of strenuous work, became possible to turn the ship into the desired direction.

The 10th of March the American icebreaker Glacier, a stronger ship than ours, tried to get through the ice, but she was not able to penetrate to our position, rather broke one of her screws and remained trapped, about 20 miles from the General San Martin, March 12th.

A period of inactivity followed because it appeared not advisable to change the position of the ship and to strain her motors only to find her trapped again at the end of the next opening. Nevertheless, the entire field of ice was moving due to wind and currents. Fortunately, this movement helped to increase the number of flaws and leads from day to day. The Glacier was able to move out by her own means on March 20, and five days later also our ship managed to get out, making good use of red flags by which the helicopters, ours as well as those of the American icebreaker Burton Island (which had remained in open waters), had marked the best leads. In the evening of March 25 we worked through the still almost compact mass of ice, and the 26th we reached our base Petrel on Dundee Island which was free of ice.

> Signed: Carlos E. Ereño
> Tte. de Navío

Courtesy of the
Armada Argentina,
SIHN, MO 4, No. 671/76
Bs As, Nov. 16, 1976

Some synoptic-meteorological comments follow: On 25 February 1975, cold, stable air moved across the southern and central Weddell Sea toward the west-northwest and thus toward the mountains of the peninsula, south of about 65°S. Due to the damming-up effect of the mountain barrier (Schwerdtfeger, 1975), a strong, northward-directed flow of cold air developed along the east side of the Peninsula. This is clearly indicated by the wind and temperature data of the stations MATIENZO (65.0°S, 60.0°W), MARAMBIO (180 kilometers to the east-northeast), PETREL (85 kilometers north-northwest from the latter), and SIGNY (South Orkneys). At the same time, relatively warm and moist air advanced southwestward over the eastern Weddell Sea; the temperature at HALLEY (75.5°S, 27°W) rose by 8° from 24 February to 26 February, to reach values 4° higher than those observed at MAT. This advection pattern suggests an intensification of the frontal zone (strong baroclinity) in the area into which a moderately developed low pressure system moved from the west (Fig. 4.18). The result was that between 25 and 26 February the direction of the flow of air in mid-troposphere turned from west-northwest to north-northwest and intensified, while the cyclone deepened. The sea level isobar pattern for 12z 26 February is shown in Fig. 4.20.

The consequences of this weather development for the icebreaker, positioned about 30 kilometers north of MAR, are obvious. The horizontal pressure gradient between the northern part of the Antarctic Peninsula and the center of the cyclone increased rapidly, and so did the wind stress on the floating ice. The variation with time of the pressure difference between MAT and SIG Island (along the thin straight line drawn in Fig. 4.18) is shown in Fig. 4.19; it must be borne in mind, however, that from these values it was possible to compute only one component of the geostrophic wind, perpendicular to the thin line.

The interpolated isobar pattern in Fig. 4.20 suggests a magnitude of the geostrophic wind vector on the order of 25 to 35 m/sec during the 36 hours of maximum storminess. That, of course, would not have created any problem without the presence of the ice.

For a short range forecast of a development like this, two features of the synoptic situation in the area 50°-80°S, 20°-90°W, might be identified; if appearing concurrently, they would warrant a 12 to 24 hour forecast of an imminent southerly storm in the northwestern Weddell Sea, and the corresponding advance of the sea ice:

i. The presence of an eastward- or east-southeastward-moving cyclone, not necessarily a strong one yet, in the region east of Tierra del Fuego or in the eastern Drake Passage. Such cyclones can easily be monitored from satellite information, the 3- or 6-hourly observations of several stations in southernmost South America and three on the South Shetland Islands, and the upper air soundings of the station BHS (62.2°S, 58.9°W).

ii. The presence of relatively high pressure south of, say, the 75°S parallel to create or maintain an easterly flow over the central Weddell Sea. As of now, only the stations HAL and BEL (77.8°S, 38.2°W) can provide the desired weather reports. Synoptic data from the essentially unexplored southwest corner of the Weddell Sea are badly needed.

4.2.5 The fateful snowstorm of December 1911

As is well known to all friends of the Antarctic, the journey of Scott and his team to the South Pole in the summer 1911/12 was affected by various kinds of difficulties, inadequate means of travel and transport, shortage of food and fuel, physical overexertion of the men, and unfriendly weather. Still, in retrospect it can be stated that most of the time the weather conditions experienced by the explorers were well within the range of the ordinary year-to-year fluctuations. The same cannot be said of what happened in the first week of December, precisely the fifth to the eighth, when Scott and his supporting parties were at 83°20'S, 170°E, about 20 km north of the foot of the Beardmore glacier. They were caught by a most extraordinary storm. In the words of Cherry-Garrard (1922): "5 Dec. 1911, we awoke this morning to a raging, howling blizzard. After a minute or two in the open one is covered from heat to foot... The temperature was +27°F this forenoon, 31 in the afternoon. 6 Dec. noon, miserable, the tempest rages with unabated violence, the temperature has gone to +33°, everything in the tent is soaking. 7 Dec. the storm continues, this day was just as warm (+35°), and wetter; our bags were like sponges. Huge drifts have covered everything. 8 Dec., wind and snow were monotonously the same, temperature +34°. Things did look really gloomy that morning." The

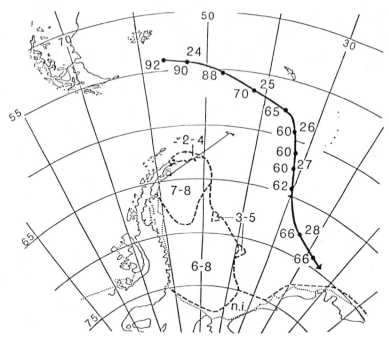

Fig. 4.18. Trajectory of the cyclone of 23-28 of February 1975. The upper numbers indicate the date, the lower ones the central pressure. 92 = 992 mb. The dashed lines delimit the ice cover of the Weddell Sea according to the Fleet Weather Facility's ice map for 27 February. Ice concentrations are given in oktas; n.i. = new ice. The thin straight line joins the met. stations SIG and MAT (see legend to Fig. 4.19).

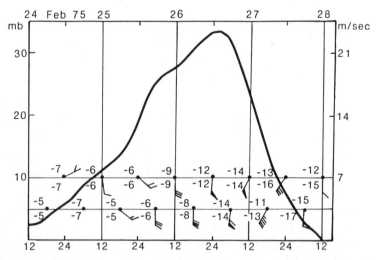

Fig. 4.19. Pressure difference (left-hand scale) and corresponding geostrophic wind component (scale on the right) between the stations SIG and MAT. Lower part: 6-hourly data of temperature, dew-point temperature and wind on board the icebreaker Gen. San Martin; 1 barb = 5 m/sec, wedge = 25 m/sec.

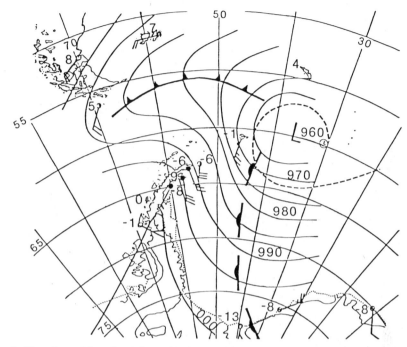

Fig. 4.20. Synoptic situation of 12z 26 February, a few hours before the storm reached maximum strength. The curved dashed line around the low pressure center indicates the position of the cyclonic vortex cloud pattern as seen from satellite. Wind symbols as in Fig. 4.19.

descriptions given by other members of the team vary only slightly: "snow like heavy wet sleet", "sometimes actual rain", "the surface 18 inches deep in slush", and so on. For three days the winds blew with gale force, and Scott called the whole event "appalling", "a raging blizzard".

From the synoptic-meteorological point of view the work of Simpson (1919) is of greatest interest. Considering the pressure and wind observations at FRAMHEIM (78.6°S, 163.6°W) and CAPE EVANS (77.6°S, 166.4°E) and the extremely high temperatures and amounts of wet precipitation at the Beardmore glacier, one can only say it was more than an ordinary blizzard. It must have been a rather large scale invasion of exceptionally warm maritime air from the northeast with considerable positive vertical motion especially on the windward side of the mountains. At FRAMHEIM, about 600 km northeast of Scott's camp near the Beardmore glacier, there was snowfall with temperatures between -0 and -2°C for three days, winds from the NE, 9 to 19 m/sec. The wind speed of 19 m/sec (hand-held anemometer) is the second strongest of all wind observations made at the station, April 1911 to January 1912. At the time of the storm, Amundsen was

already on the high plateau, at about 3000 m elevation, 1000 km south of
Framheim, 500 km southsoutheast of Scott's camp; he experienced exceptionally
bad weather conditions too (observations cited in detail on pages 58-59).
Simpson (ℓ.c., p. 308) who at that time was at CAPE EVANS noted at 8 December:
"Two days previously (i.e. 6 Dec.) there had been the greatest snowfall of the
year, which had been in the unusual form of large flakes. The whole of the
frozen McMurdo Sound was covered with about 18 inches (nearly 50 cm) of light
loose snow." A similar type of storm, only in a more appropriate part of the
year, is mentioned by Mitchell (1970): "As a tongue of warm moist air is advected into the southern Ross Sea and across the ice shelf, rapid (almost explo-
sive) deepening of the semi-permanent Ross Sea low may occur. On one occasion,
April 1967, Byrd Station reported a 26°C temperature rise in 24 hours. At the
end of this period,... the pressure at McMurdo had dropped to 951 mb."

For Scott's march to the Pole, the snowstorm at the foot of the
Beardmore glacier was a catastrophic event. It meant the loss of four full
summer days with all possible consequences regarding the tightly rationed supplies to be carried south. Furthermore, during the first days after the snowstorm when the party had still to reach the foot of the glacier and begin the
ascent to the high plateau, the surface conditions could not be worse; it
required superhuman efforts to move ahead, not more than a few miles per day.
One can only sense awe, or feel compassion, when reading the individual reports.
It is a fair guess that without the loss of these days, i.e., without interven-
tion of this extraordinary weather phenomenon, Scott and his men would have
made it home.

4.2.6 Blocking and other anticyclones

The term "blocking anticyclone" is generally used in synoptic meteorology to describe tropospheric high pressure systems that remain approximately
stationary for several days in an area where zonal flow prevails and the frequent passage of cyclones is the norm. In the middle and subpolar latitudes of
the Northern Hemisphere the phenomenon has been well investigated in context
with the study of the wave pattern of the atmospheric circulation. For the
southern ocean we can say that blocking anticyclones are certainly of shorter
duration and, at least south of 50°S, rather rare events. Still, over the open
ocean and in the surroundings of small islands, like some of the South
Shetlands, such events can be of practical interest. In fact, they are the only
kind of synoptic situation with a high probability of a few days of weak, or
not more than moderate, surface winds, welcome for oceanographic work, disembarking on islands, and other outdoor activities.

The example to be briefly discussed was chosen in spite of having occurred in the pre-IGY, pre-satellite days, because it really is an extraordinary case showing what <u>can</u> happen in a region well known for its storminess. As described by Grandoso and Nuñez (1955), an anticyclone remained for a period of ten days centered between Falkland and South Shetland Islands (Fig. 4.21). From the daily soundings made at PORT STANLEY Table 4.8, it can be seen that through the entire troposphere the temperatures were above normal, and the prevailing winds blew from the northeast. These conditions characterize a high-reaching anticyclone with its center in the middle and upper troposphere southeast of the Falklands. Any transport of air masses from west to east is evidently blocked; hence it is important for practical as well as dynamic meteorology.

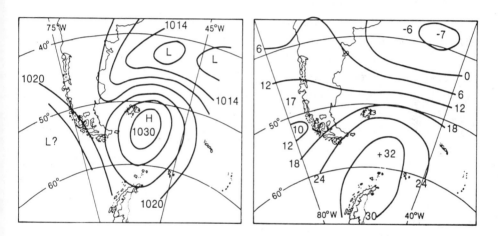

Fig. 4.21. A blocking situation east of the Drake Passage, after Grandozo and Nuñez (1955). At left: average sea level isobars (3 mb increments) for the ten days 5-14 June 1952. At right: isolines of deviation of this average from the multiannual mean chart for June (6 mb increments).

TABLE 4.8. Resultant wind and mean temperature above PORT STANLEY (51.7°S, 57.8°W), 4-13 June 1952, compared with 9-year average values for June.

Period	Level	700	500	300	200 mb	
4-13 June	D_R and V_R	062°8	053°12	043°16	007°10	m/sec
number of soundings		10	10	9	8	
9 years June	D_R and V_R	265°12	263°16	266°22	267°20	m/sec
4-13 June	T	- 9	-26	x	-63	
9 years June	T	-12	-30	x	-60	

A different type of anticyclone can form in the realm of the cold air which moves toward north or northeast on the back side of eastward progressing cyclones. The frequency of occurrence attributed to this type of migrating, low-tropospheric anticyclone depends on the minimum conditions which a pattern of isobars is expected to fulfill to justify the term "anticyclone". Internationally recognized definitions do not exist. Still, for the southern ocean south of 60°S, high pressure systems which remain identifiable through several days are infrequent phenomena, as can be seen in the maps of anticyclone tracks south of 30°S during the IGY (Taljaard and van Loon, 1962 and 1963; Taljaard 1972). These maps are based on the daily sea level weather charts July 1957 - December 1958, elaborated by the South African Weather Bureau (1962-1966).

The weather conditions in such pressure systems depend mostly on the character of the surface, -- 10/10 sea ice, fractional pack ice cover, or open ocean waters, -- and the stability of the original air mass in the lower troposphere. More about this in Chapter 5. The weather on the coast of the continent itself, in the migrating anticyclonic ridges between two coast-parallel passing cyclones, is generally controlled by the topography of the hinterland more than by the larger scale circulation pattern.

4.2.7 Fronts, air masses, and jetstreams

On the high plateau of Antarctica, the temperature in the lowest few meters of the atmosphere can change drastically, for instance when the radiation inversion near the surface is destroyed by rapidly increasing winds. In the coastal regions, pronounced variations of temperature and moisture content of the air in the surface boundary layer occur when open water regions are not far away and the surface wind direction suddenly varies from seaward to landward or vice versa. In many such cases, one should not immediately conclude that a "front" has passed; it is possible, but no more than that. In synoptic meteorology a "front" is defined as a sloped, shallow transition layer (in theory idealized into a two-dimensional frontal "surface") between two different air masses whose vertical extent amounts to a few kilometers, not a few meters or tens of meters (Taljaard et al., 1961). It can happen, of course, that a pronounced surface temperature change is accompanied, preceded, or soon followed by a radical change of cloud type or amount, or of precipitation; then a real front is easy to recognize. In some cases, however, a front can be identified only when upper air soundings are available, and almost always the effect of a front on the weather development could be estimated much better if rawinsoundings were carried out in shorter time intervals than the 12 or 24 hours generally applied, but that is generally not feasible for understandable logistic reasons.

So it is that in the past 30 years there has been relatively little research regarding the characteristics of fronts in the Antarctic, certainly not enough to doubt the reasonable assumption that fronts in the southern polar regions do not differ essentially from those in the Arctic. This statement refers in particular to aspects like the slope of the transition layers or frontal surfaces, their modification over mountainous terrain, destabilization due to ageostrophic flow in middle and upper troposphere, and details of the occlusion process.

The contrast in temperature, moisture content, and wind between two air masses can be particularly large when the front or transition layer between them is not far from the coast. The surface water temperature changes relatively little with the seasons and with latitude; Table 4.9 shows a few mean values.

TABLE 4.9. Average surface water temperature for February and August in two sectors of the southern ocean, at 50 and 60°S latitude; evaluated from the sea surface temperature charts of the Southern Ocean Atlas (Gordon et al., 1982). The asterisk marks an estimated value; between the pack ice, the water temperature is near -1.8°.

| | Pacific Ocean 160°E - 180° - 80°W | | Atlantic and Indian Ocean 60°W - 0° - 100°E | |
	FEB.	AUG.	FEB.	AUG.
50°S	10	7	6	3°
60°S	4	1	1	-1^*

The average surface _air_ temperatures over the southern ocean vary correspondingly little. In contrast, the temperature differences between February (or March) and August (or September) over land or ice are much greater; many examples are given in Table A3. Furthermore, air moving toward higher latitudes will generally be cooled from below, a process which increases vertical stability, so that convection cannot reach high. Therefore, maritime polar air (MP) can easily be identified when upper air soundings are available. At any height or pressure level above the boundary layer the values of temperature and specific humidity vary within modest limits.

This conservative character of maritime polar air in the free atmosphere is tempered only by the long wave radiative loss of heat which, on the average, causes a cooling between 1 and 2° per day. Therefore, even in the highest latitudes, in winter more than 2000 km distant from the ice-free ocean waters, maritime polar air can be identified. Fig. 4.22 gives an example of such an air mass at the SOUTH POLE, 14 August 1974, advected from lower to higher latitudes in the preceding days. The Figure also shows the invasion of continental antarctic air, coming along the 90°E meridian from the central, highest and coldest part of the Plateau.

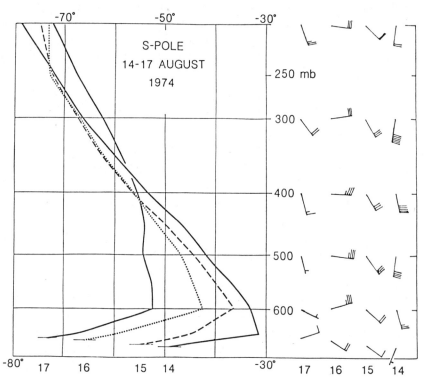

Fig. 4.22. A change of air masses from maritime polar to continental antarctic air. On the 14th, the winds are blowing toward the Pole along the meridian 180°, on the 16th along 90°E. A long barb indicates 5 m/sec.

Maritime antarctic air (MA) might be the name for air masses that have stayed, generally at higher latitude than MP, over sea ice, ice shelves, or snow-covered land of minor elevation, long enough to assume a low-level inversion profile typical of their environment; both temperature and moisture content are considerably lower than in MP, particularly in the eight or nine colder months of the year (Dalrymple, 1966). The range of characteristic temperatures for MA should be larger than for MP, due to the less uniform surface conditions and smaller possible source area.

Finally, the name continental antarctic air (CA) should belong to airmasses that have adapted, or come near to, a zero energy balance at surface and in the free atmosphere over the high plateau, above, say, the 2000 m level. The area is certainly large enough (> $8 \cdot 10^6 km^2$) to make such an adjustment possible, an aspect which could remain in doubt for the air masses called MA. The characteristic vertical profiles of CA, monthly or seasonal, must indicate lower temperature and smaller specific humidity values than the corresponding monthly or

seasonal average profiles of the other two air masses. When analyzing a situation in which a relatively warm and moist air advances poleward, one should be
aware that the change of the air masses over a given place or region does not
necessarily require the passage of a sharp front. It also can be a smooth and
steady transition.

When a sizeable contrast between two air masses extends into the upper
troposphere, the slope of the isobaric surfaces (rising toward the warmer air)
and with it the wind increase strongly with height, sometimes drastically.
Since the original work on "Strahlstroeme" (i.e. jet streams) by Seilkopf (1939)
and the much better known and wider reaching research after World War II at the
Chicago School mainly under Rossby and Petterssen (Reiter, 1964), location and
strength of the jet streams in the upper troposphere and their relation to
cyclogenesis are essential ingredients to the reasoning of analysts, forecasters or programmers in synoptic meteorology. Of course, all this has been
developed in, and with reference to, the extratropical Northern Hemisphere,
where upper air soundings of a network of weather stations as well as air
reports made by "jet airplanes", all in combination with the satellite pictures
and data, make that feasible.

The supply of upper air information in the Southern Hemisphere is much
more modest, especially between 50° and 65°S, as explained in Chapter 1. In the
broad band of maximum cyclonic activity over the southern ocean the only help
is satellite information. Whenever pictures in the visible and infrared are
available, the approximate location and orientation of the jet streams can
often be seen. Numerical values, however, for strength, length, and width of
the maximum wind field, perhaps 2000 km long in wind direction but not more
than 300 km extended transversely, can at best be guesswork. They do not suffice to quantitatively determine divergence and vorticity and compute vertical
motions in the environment of the jet streams where such parameters are particularly important for a correct, three-dimensional weather analysis.

It stands to reason that one sounding per day at one place in an area of
a million km^2 cannot help much for the daily hemispheric upper air analysis.
Nevertheless, much can be learned from the rawinsoundings carried out in the
years 1962-66 aboard the USNS research ship Eltanin operating in the South
Atlantic and South Pacific (ESSA, 1968). Table 4.10 shows the frequency of the
occurrence of jet stream speeds -- the name is generally used when the wind
speed > 30 m/sec -- over that part of the southern ocean. The number of days
with such winds is probably an underestimate; days with windstorms near the surface, which make the launching of sounding balloons practically impossible,
often are also days with very strong upper winds. Table 4.10 compares these
data with those at a coastal station at 69°S and two inland stations, at 80°

TABLE 4.10. Frequency of wind speeds ≥ 30, 40, and 50 m/sec at the 300 mb level, measured aboard USNS <u>Eltanin</u> in the South Atlantic and South Pacific south of 50°S, and at three stations at 69, 80, and 90°S, in the winter half year. Evaluated from extensive publications: ELTANIN: ESSA (1968) Vol. 9, for 1963-1966 and 1968; SYOWA: Japanese Met. Agency, Antarctic Data for 1973-1977; BYRD: ESSA (1970) Vols. 10 and 11, for 1967-1969, and unpubl. data 1965-1966; S-POLE: ESSA (1970) Vols. 11 to 14, for 1969-1975.
N = total number of wind soundings reaching or exceeding the 300 mb level,
n = number of soundings with indicated speeds,

Station	N	≥ 30		≥ 40		≥ 50 m/sec		maximum	
		n	%	n	%	n	%	m/sec	from
ELTANIN	292	149	51	80	28	33	11	83	276°
								at 61°S, 95°W	
SYOWA	1620*	185	11	58	4	17	2	80	208°
								see Fig. 4.23	
BYRD	821	111	14	28	3	6	0.7	57	324°
S-Pole	1095	53	5	7	0.6	2	0.2	53	156°

*)Twice daily soundings since 1974

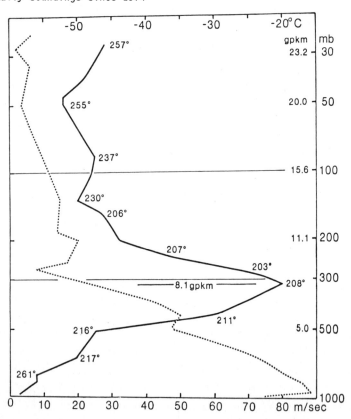

Fig. 4.23. A strong jet stream over SYOWA Station (69°S, 40°E), 00z 5 April 1974. Solid line= wind speed, scale at the bottom; three-digit numbers = wind direction; dotted line = temperature, scale at the top of the graph. Along the right-hand side ordinate the height scale in mb and in (geopotential) kilometers.

and 90°S. The 300 mb level has been chosen as reference level throughout; when jet streams are found in the upper troposphere of the polar and subpolar regions, with few exceptions the level of maximum wind is not far from 300 mb. No further discussion of the context of the table appears to be needed. Still, the example of a very strong jet stream in Fig. 4.23 might help to visualize what can happen in the free atmosphere even when the surface observations could suggest a tranquil day; there were only a few (< 1/10) altocumulus clouds in the sky, the visibility 50 km, the wind weak, and the pressure rising moderately. There are other possibilities. Fig. 4.24 shows a situation with very strong winds again, but in this case the term "jet stream" would not be appropriate. Rather, within no more than twelve hours, the entire troposphere, and probably a major part of the stratosphere too, changes from a normal winter temperature

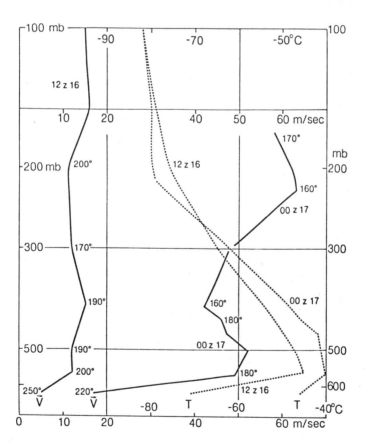

Fig. 4.24. Fast development of extremely strong winds in the troposphere and weakening of the surface inversion at Vostok, 12z to 24z, 16 September 1965. The wind speed is given by the two solid lines, the direction in numbers, the temperature by the two dotted lines; 100 mb ≈ 14 km, approx.

profile and winds < 15 m/sec into a different profile with warm air below 9 km height, blowing snow at the bottom and winds beyond 40 m/sec as far as the measurement reaches.

4.3 ATMOSPHERIC PRESSURE SYSTEMS OVER THE ANTARCTIC PLATEAU
4.3.1 Problems of terrain elevation and pressure reduction

Any analysis of the pressure field, in the form of isobars at different height levels or contour lines of constant pressure surfaces, above the higher (> 2500 m) part of the continent involves various difficulties which since the IGY have not always received the attention they deserve. In the interior there are at present only two stations, SOUTH POLE and VOSTOK, with daily surface and upper air measurements. Until 1970 also BYRD Station, near the 1500 m level in West Antarctica, was in continuous operation. South of 71°S latitude and not far from sea level, there are two more stations with upper air soundings, McMURDO and HALLEY; a third one, ELLSWORTH, was alive until 1963.

The distance from the SOUTH POLE to VOSTOK is 1300 km, to MCMURDO 1350 km, to HALLEY 1600 km. All this makes it illusory to look at any level for clearly defined and accurately located cyclones or anticyclones, troughs or ridges. In many cases they might exist, though one cannot prove it because small but not necessarily weak pressure systems with a characteristic radius < 1000 km are entirely possible.

To determine the geopotential height of any pressure level in the atmosphere, the elevation of each upper air sounding station must be known. That is no problem for the coastal stations. A height increase of 8 m approximately corresponds to a pressure decrease by 1 mb, and for most practical purposes an exactness of ± 0.5 mb is sufficient. For the two stations in the interior of the continent, the situation is quite different. When regular meteorological measurements at Amundsen-Scott Station (SOUTH POLE) were initiated in 1957, the height above mean sea level of the site was not known with sufficient accuracy. An approximate value of 2800 m was assumed, corresponding to 2808 gepot. meters. In the following 18 years, this value was employed for pressure-height computations, in spite of the fact that in the early 1970s the elevation of the snow surface at the South Pole was determined to be close to 2846 m. That value was introduced, without any further comment or explanation but with unrealistic accuracy, as 9340.6 feet in the station description of the NOAA publication "Climatological Data for Amundsen-Scott, Antarctica", for 1974 and 1975, No. 14, June 1977. Unfortunately, in the other much used periodical published by NOAA, "Monthly Climatic Data for the World", no notice was taken of the change. Nevertheless, the value of 2846 m has been employed since 1 January 1975 for the evaluation of the daily upper air soundings at the S-POLE Station itself.

Finally, after these and other discrepancies had come to the attention of researchers, the U.S. Geological Survey through W.H. Chapman made a new evaluation of the Geoceiver information accumulated since 1972. More details are given in a note in the Antarctic Journal of the U.S. (Schwerdtfeger, 1981). The result "within 5 m exactness" is the following: 2835 m (2843 gpm) is the elevation of the snow surface at the South Pole which should be used and, to the author's knowledge, has been used for the computation of the daily soundings since 1 January 1982. Corrections to be applied to all earlier published geopotential height values obtained by radiosondes released at the Pole are: +35 gpm for 1957-1974; - 12 gpm for 1975-1981.

Only a short remark shall be made regarding the other upper air station in the interior. The elevation of VOSTOK is 3488 m, according to Averyanov (1972), Dolgina et al. (1977) and other Russian publications. Nothing is known (to the author) about the accuracy of this value, but it can certainly be assumed that the lower value of 3420 m, which had appeared in Russian publications prior to 1968, is not tenable. Unfortunately, still in 1982 this lower value is to be found in the NOAA-WMO publication "Monthly Climatic Data for the World" whenever "CLIMAT" data for VOSTOK are reported.

Finally, the custom of some institutions to draw sea level isobars over an area of several millions of km^2 with more than 3000 m elevation needs to be mentioned. The author is in no position to justify the reduction of a pressure value measured near the surface to a pressure at sea level when the height difference is so large. The point is that the computation requires an arbitrary assumption of a mean temperature of a vertical column from one to the other level. For the SOUTH POLE Station, for instance, a change of that temperature by only 1° leads to a change of the fictitious sea level pressure by 1.4 mb, for VOSTOK 1.7 mb, and the uncertainty of the chosen temperature is certainly a multiple of one degree. Knowledge of the mean temperature of a column of firn, ice, and rock does not help; an atmospheric layer would react quite differently to a real change of pressure and temperature at surface. One must conclude that sea level isobars on the antarctic plateau do not have a clear physical meaning. It is hoped they will disappear soon.

4.3.2 Vertical thermal structure related to surface pressure

The thickness of a layer of air between two constant pressure surfaces is inversely proportional to the density of the air and thus directly proportional to the temperature. It follows that the slope of isobaric surfaces representing the intensity of cyclonic pressure systems can increase with height only when the air in the central part of the cyclone is colder than the air around it. Analogous reasoning applies to anticyclonic pressure systems, intensifying with height when there is relatively __warm__ air in the center. These basic rules, hotly debated by our professional ancestors about 100 years ago (Hann, 1891), can help one to understand that over the high plateau of Antarctica the tropospheric air in the low pressure systems is cold, in high pressure systems comparatively warm. Over the southern ocean and over low-lying land or ice, warm cyclones and cold anticyclones weaken with height to insignificance in the lowest three kilometers. Only pressure systems of the opposite structure can persist and remain recognizable when they move from the subpolar latitudes toward and across the main part of the continent.

The predominant appearance of cold lows and warm highs becomes evident in Fig. 4.25 a and b, showing the relation between surface pressure (p_s) and thermal structure of troposphere and lower stratosphere at SOUTH POLE in winter and summer: on the right-hand side two examples of extremely low and extremely high p_s, on the left the average thermal structure for days with p_s considerably below and days with p_s above average. In the latter graph, each group represents about one third of the soundings launched in 1971 to 1975, reaching or exceeding the 100 mb level.

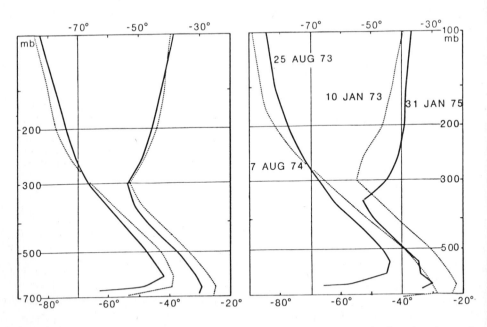

Fig. 4.25. The relation between surface pressure and vertical structure of troposphere and lower stratosphere at the South Pole: cold low and warm high. a) Average 1971-75, July and August: total number of days with soundings = 296 ≈ 95% of all days; of these, 99 days with p_s < 673 mb and 95 days with p_s > 684 mb. December and January: total number of days = 285 ≈ 92%; 80 days with p_s < 687 mb and 82 days with p_s > 695 mb. b) On the right-hand side: extreme cases.

Inverse temperature conditions impose themselves in the stratosphere, so that the thickness of the 600 to 100 mb layer is approximately the same for the 95 days (of a total of 296 days with soundings in July and August 1971-75) with p_s < 673 mb, as it is for the 100 days with p_s > 684 mb, 10,820 and 10,830 gpm respectively. This carries the noteworthy implication that the advection of cold or warm air masses in the shallow layer (generally < 1 km) between surface and the 600 mb level, and also the atmosphere above the 100 mb (about 15 km) level, must be of great importance for the change (with time) of the pressure at surface. One might also speculate that pressure systems (lows and troughs, highs and ridges) over the antarctic plateau are advected, not generated in situ.

The coherence of the temperature in the warmest tropospheric layer and the pressure at station level can be of interest for ice crystal formation and precipitation, discussed in the following chapter. For the SOUTH POLE, the 600 mb level has been chosen to represent the warmest layer. Table 4.11 contains statistics from the same data set as used for Fig. 4.24, only this time taking three groups of approximately equal size according to the temperature at 600 mb.

TABLE 4.11. Temperature in the warmest layer and pressure at surface, South Pole, July and August, 1971-75.

Temperature group	<- 41°	-41 to -38°	> -38°
Number of soundings	99	104	103
Average temperature	-43.8	-39.4	-34.9°
Average sfc pressure (p_s)	671	678	685 mb
Standard deviation of p_s	7.5	9.2	8.4 mb
Prevailing type (p_s)	cold low	near average	warm high

Understandably, in the quiet summer season the contrasts are smaller than in the winter, but still large enough to make it possible that the troposphere above the boundary-layer on a day of August with exceptionally high surface pressure can be as warm as on a January day with exceptionally low surface pressure. The main characteristics, cold low pressure and warm high pressure systems in the entire troposphere, are essentially the same all year round.

Chapter 5

H_2O, AS GAS, LIQUID, AND SOLID

H_2O in the vapor phase is always a part of the mixture of gases called air. There is no such a thing as "absolutely dry air" in Nature, though the concentration of water vapor in air can vary by a wide range, from about 4×10^{-2} [kg of water vapor per kg of air] in extremely humid tropical air to 5×10^{-3} in maritime subpolar air to less than 1×10^{-6} on the high plateau at cold days of the winter half year. When changes of phase occur from gas to liquid (condensation), from liquid to solid (freezing), or from gas to solid (deposition), energy in the form of latent heat is released. Correspondingly, energy supply (by absorption of radiative energy or by conduction of heat from the surrounding air) is needed for the opposite processes, evaporation, melting, and sublimation. Since the H_2O content of antarctic air is so small, the amounts of energy released or required are not as large as in warmer air masses, and the saturation-adiabatic temperature lapse rate differs but little (not more than by 10%) from the dry-adiabatic (9.7°/km). Nevertheless, without consideration of these six thermodynamic processes it would be impossible to understand what is going on at and above the surface.

The formation and accumulation of snow and ice must be the most important process in the Antarctic. Great amounts of solid H_2O are moving continuously from the continent toward and into the southern ocean. Similar amounts have to be transported to, and deposited on, the continent in order to maintain the approximate balance that has existed for many thousands of years. The question of how this transport comes about and what conclusions can be drawn regarding the distribution of precipitation of snow on the continent will be discussed in Section 5.2. First, a few remarks will be made about the results of measurements of atmospheric humidity, because even in the era of satellites and computers such measurements present serious difficulties for practical meteorology in polar regions.

5.1 THE WATER VAPOR CONTENT OF POLAR AIR

Several parameters can be used to quantitatively indicate the water vapor content of the air directly or indirectly: partial pressure of the vapor (briefly "vapor pressure"), its density, mixing ratio, specific humidity, relative humidity, and dew-point or frost-point temperature. In the following, mostly the terms vapor pressure and specific humidity will be used, the latter defined as the ratio "mass of water vapor per mass of air" = "density of water vapor per density of air", kg of H_2O/kg of air. Since this value is generally small, in practical meteorology one prefers to use the ratio <u>gram</u> H_2O per kg air. In polar air masses it is irrelevant whether one uses the specific humidity or the mixing ratio, the latter defined as the ratio "mass of water vapor per mass of <u>dry</u> air". For instance, the value of the specific humidity at temperatures near 0° and pressure near 1000 mb would have to be multiplied by 1.004 (at 500 mb 1.008) to get the corresponding mixing ratio; at -20°, the factor would be 1.0008, (1.0016). In polar regions up to now, no series of measurements have ever been made whose accuracy would justify to take the resulting differences into account.

The state of equilibrium between liquid water and water vapor, i.e., when there is no net loss or gain for one or the other medium by evaporation or condensation, is called saturation. To be precise, the numerical values of the saturation vapor pressure refer to a plane surface of pure water. They are a function of temperatures only. In consequence, saturation specific humidity must depend on temperature <u>and</u> pressure. Saturation values as shown in Table 5.1 are not absolute limits of the possible water vapor content of a mass of air. Supersaturation is possible. With respect to water it amounts only to very small quantities, certainly less than 1% of the total mass of water vapor per mass of air. In contrast, under conditions as found in the interior of Antarcica, supersaturation with respect to ice can be much larger, possibly by 20% or more, as a persistent state of the atmosphere in suitable layers (more about this in Section 5.4).

TABLE 5.1. Saturation vapor pressure e_s and saturation specific humidity q_s (at 1000 and 500 mb) as function of temperature; w stands for "over water", i for "over ice". Complete tables can be found in the Smithsonian Met. Tables (List, 1959).

T	+10	0	-10	-20	-30	-40	-50°	
e_{sw}	12.27	6.11	2.86	1.25	0.51	0.19	---	mb
e_{si}	---	6.11	2.60	1.03	0.38	0.13	0.04	mb
e_{si}/e_{sw}	---	1	.91	.82	.75	.68	.62	--

at p = 1000 mb

	+10	0	-10	-20	-30	-40	-50°	
q_{sw}	7.70	3.82	1.79	0.78	0.32	0.12	---	gr/kg
q_{si}	---	3.82	1.62	0.65	0.24	0.08	0.025	gr/kg

at 500 mb

	+10	0	-10	-20	-30	-40	-50°	
q_{sw}	15.45	7.65	3.58	1.57	0.63	0.24	---	gr/kg
q_{si}	---	7.65	3.24	1.29	0.47	0.16	0.049	gr/kg

The relationship between temperature and saturation vapor pressure shown in the upper part of Table 5.1 can be derived from the laws of classical thermodynamics and be expressed in the form of the so-called Clausius-Clapeyron equation (Iribarne and Godson, 1973). In the real atmosphere the state of saturation with respect to water is found with good approximation in thick water clouds or dense fog. Under any other conditions the air tends to be unsaturated. In the context of clouds or fog, an important characteristic of moist air has to be mentioned. Liquid water in form of droplets can persist in the atmosphere in the state of being supercooled, i.e., at temperatures below freezing, even when there is a continuous loss of heat. On the other hand, ice particles floating in the air start melting at temperatures above 0° when the surrounding air can serve as heat source.

Some seasonal averages of the actual specific humidity computed from the three-hourly measurements about 1.5 m above the surface are shown in Table 5.2. Such values cannot claim a high degree of accuracy. Inadequate functioning of different types of instruments and inclement weather conditions impeding precise measurements combine to impose a margin of error which increases as the temperature of the air decreases below the freezing point. Also, for some stations the disregard of the non-linearity of the relation between T and e_s has contributed to the inaccuracy of daily and monthly values. Neither the average of a series of relative humidity values (actual mixing ratio per saturation mixing ratio times 100) nor the average of a series of dew-point temperature (the temperature to which the air would have to be cooled to reach the state of saturation) permit the derivation of the average specific humidity or mixing ratio (Schwerdtfeger, 1970a, p. 296).

TABLE 5.2. Seasonal mean values of the specific humidity (in grams of water vapor/kg of air) near the surface.

Station[*]	Lat.	Dec.+Jan.	Feb.+Mar.	Apr.-Sep.	Oct.+Nov.	Years
VOSTOK	78.5°S	0.29	0.07	<0.01	0.07	1958-69
HALLEY	75.5	2.4	1.4	0.6	1.2	1971-81
SANAE	70.3	2.3	1.3	0.5	1.0	1971-81
MOLO	67.7	2.3	1.6	0.9	1.3	1963-72
CASEY	66.3	2.8	2.1	1.1	1.6	1959-67 1976-81
FARADAY	65.3	3.4	3.3	2.1	2.7	1972-81
ORCADAS	60.7	3.2	3.3	1.8	2.7	1906-50

[*]for U.S. stations in the Antarctic, moisture values have been published only for the radiosonde ascents.

One of the advantages of using the specific humidity or the mixing ratio to indicate the moisture in the air is that it remains constant when the temperature varies, i.e., when heat is added to or withdrawn from the air, provided neither evaporation nor condensation take place and the change of pressure p remains insignificant (regarding the latter, see the slightly sloped lines in Fig. 5.1). Therefore, even without great accuracy of the measurements, a comparison of the saturation values in Table 5.1 with the averages of actual specific humidity in Table 5.2 points to a biological problem which becomes important for the coastal stations mostly in the winter, in the interior of the continent at any time: when the air in a building is heated to a tolerable indoor temperature without an adequate supply of water vapor, the relative humidity gets very low, < 10% on an average winter day at HAL or SANAE, < 1% at S-POLE or VOSTOK, certainly far below the lower limit for well-being of the human body which would lose a considerable amount of H_2O with every breath. The primitive remedy is a pot of boiling water on the stove, and a cup of water (or whatever else) in the hand.

After a look at the water vapor content of the atmosphere's lowest layer, the next question must be how it changes with height. The answer is summarized in Fig. 5.1. The earlier statements regarding doubtful accuracy apply also here. Nevertheless, the valuable soundings made aboard the research ship Eltanin under the arduous conditions of the stormy southern ocean, and the daily soundings of three east antarctic coastal stations give altogether a clear picture. The moisture content decreases with height in the lower half of the troposphere, and there is no doubt that it continues in the upper half. The general decrease poleward is also in evidence, with some longitudinal variations in agreement with the circulation pattern, in the sense that more frequently there are winds with southerly component (dryer air) over MAWSON than WILKES (CASEY) and SYOWA.

Over the high plateau the situation is quite different. There are not any reliable series of measurements, but there are good reasons, explained in Section 5.4, to assume that during most of the year the air in the layer below 5 km (m.s.ℓ.) is saturated or supersaturated with respect to ice. Reliable temperature soundings of SOUTH POLE and VOSTOK are available for many years. Therefore the curves for these two stations in Fig. 5.1 have been drawn for ice-saturation specific humidity; in the upper part of the troposphere the real values would probably be a bit smaller still. The profiles for July, qualitatively representative at least for the eight months March to October, show an increase of the water vapor content from the cold surface to the much warmer layer in and above the inversion. In the short summer, there appears to be a significant difference between the profiles for SOUTH POLE and VOSTOK. At the latter place, $11\frac{1}{2}°$ latitude off the Pole, the diurnal variation of the surface

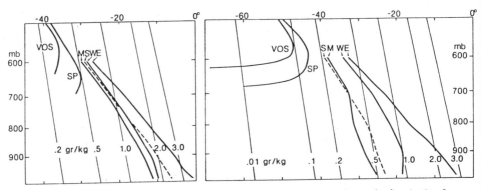

Fig. 5.1. Average water vapor content (specific humidity in gr/kg) of the lower half of the troposphere over the southern ocean between 60 and 65°S, and three coastal stations, from radiosonde ascents; summer data at the left. Data: research ship Eltanin, 40 soundings in December and January 1963-66, and 64 soundings June-August 1962-66; MAWSON and WILKES (CASEY), January and July daily soundings 1965-68, SYOWA 1970-73. In the upper left of the two graphs, saturation specific humidity with respect to ice in January and July at SOUTH POLE and VOSTOK. For these two stations only the pressure and the respective temperatures used (multiannual monthly means) are based on measurements; the state of ice saturation is assumed. On most winter days some supersaturation with respect to ice probably exists. The thin, slightly sloped lines are lines of constant specific humidity.

temperature is large (as it is at PLATEAU Station, Fig. 2.6, p. 32) and the vertical exchange of heat and moisture is obviously much stronger than at the Pole where the periodic diurnal variation of T is zero. Hence, the increase of moisture content with height in mid-summer will be periodically interrupted over most of the area of the Plateau, weakly remain only at the Pole and proximity.

Based on data as used in Fig. 5.1, it is easy to determine approximate values of the total amount of water vapor in a vertical column (over a horizontal reference area) extending from the surface up to a level above which the water vapor content of the air is insignificant. What can be considered as "insignificant" depends, of course, on the requirements of the particular investigation as well as the possible exactness of the measurements; for polar air, the expectations should not be set too high. The name of the quantity found in most textbooks, "precipitable water", is misleading, to say the least. If all water vapor could be brought to condensation or deposition and then precipitate, absolutely dry air would remain; that is not the way Nature operates. The more appropriate term "liquid equivalent" has not yet been generally accepted. The original mathematical formulation and the summation formula used in practice are the following:

$$LE = \int_{p_u}^{p_s} \frac{\rho_v}{g\rho} \, dp \doteq \frac{1}{g} \int_{p_u}^{p_s} q \, dp \doteq \frac{1}{g} \Sigma \, \bar{q}_i \, \Delta p_i \; [kg \; H_2O \cdot m^{-2}] \tag{5.1}$$

with p_s and p_u the pressure at surface and at an upper level, g = gravity, ρ_v and ρ = densities of water vapor and of air, q = specific humidity, Δp_i = difference between the pressure at the lower and upper level of an atmospheric layer i; in practice, Δp might be 50 or 100 mb. Values of LE at four stations between 60 and 90°S are shown in Table 5.3. A typical value for moist tropical air would be 50 kg $H_2O \cdot m^{-2}$.

TABLE 5.3. Approximate average values of the liquid equivalent of the water vapor content or "precipitable water" in a vertical column from surface to tropopause, at various antarctic stations in summer and winter. 1 kg H_2O/m^2 is equivalent to 1 mm precipitation.

	SOUTH POLE 1971	L. AMERICA 1958	WILKES 1958	ORCADAS 1958
JANUARY	1.6	4.8	6.9	9.4 kg H_2O/m^2
JULY	0.3	1.3	3.3	4.4 kg H_2O/m^2

The values for SOUTH POLE have been computed assuming saturation with respect to ice. For the other three places, values of the relative humidity U = q/q_s × 100 were measured and published up to the 500 mb level; the small contribution of the upper layer has been extrapolated.

5.2 THE TRANSPORT OF H_2O POLEWARD ACROSS THE COASTLINE

The following considerations, based mostly on the work of D. Bromwich (1978, 1979), will refer to East Antarctica. The reasons are that only for that part of the continent a simple but not yet too unrealistic model can be used which substitutes a parallel for the real coastline, and that only between 5°W and 115°E, one-third of the continent's total circumference, there is a chain of nine meteorological stations with upper-air soundings.

It is a hopeless enterprise to measure by conventional methods the precipitation falling on the coast and the first few hundred kilometers inland, where the terrain rises steeply and irregularly and strong surface winds are the rule rather than an exception. On the other hand, the very small amounts of precipitation which fall on the high plateau of the interior are rather well known, and even an error of ± 20% of the mean annual values would make a difference of less than ± 1 gram/cm^2. Therefore, it has been examined if the meridional transport values allow to estimate the amount of H_2O available for precipitation on the continent, and in particular on the 300 to 500 km wide band adjacent to the coast. The assumption had to be made that there is no net contribution of H_2O by zonal transports.

5.2.1 Meridional water vapor transport

The first step is to determine the total mass of water vapor which at a given location moves (south- or northward) in a unit of time (a month, for instance) across a unit of length on the parallel representing the coastline. In other words, one is looking for the time average of the vertically integrated meridional transport of water vapor, Q, as defined in equation (5.2),

$$\overline{Q} = \frac{1}{g} \int_{p_u}^{p_s} q\,v\,dp = \frac{1}{g} \int_{p_u}^{\overline{p}_s} \overline{q}\,\overline{v}\,dp + \frac{1}{g} \int_{p_u}^{\overline{p}_s} \overline{q'v'}\,dp \left[\frac{kg\ H_2O}{m\ sec} \right] \quad (5.2)$$

where the notation is the same as in equation (5.1), and v = meridional wind component, positive when from the south; the overbars indicate time averages, the primes stand for departure from the mean. The two additive terms on the right-hand-side of the equation represent the contributions of the mean flow and the eddy flux of water vapor, respectively. This kind of approach is possible, of course, only when daily, and better twice daily, soundings are available. Fig. 5.2 and Table 5.4 show the importance of the eddy term.

On the way to obtaining unbiased transport values there are at least two pitfalls which were not recognized, or at least not taken into account, in earlier studies. The first source of error lies in the vertical spacing of the levels for which the wind and the moisture content of the air have been evaluated from the radiosonde records and used for the computation of $\overline{Q}$. The second type of errors comes about by disregarding that at most stations there are days without soundings.

Fig. 5.2, taken from Bromwich (1978), explains that only detailed vertical profiles of v and q for the lowest 1000 or 1500 m can lead to realistic transport values. A comparison of the straight, dashed line from surface to the 850 mb level with the dotted line leaves no doubt that profiles produced with only the data usually published in station annals lead to a sizeable overestimate of the outflow from the continent. Several windspeed profiles shown in another context in Fig. 3.10 (p. 68) confirm this statement. The layer of cold

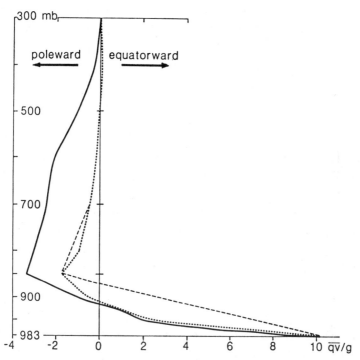

Fig. 5.2. Profiles of the meridional transport of water vapor at MIRNY, August 1972, 00 and 12z (after Bromwich, 1978). The solid line represents the total transport term $\overline{q\,v}/g$, the dotted one the term computed only with mean values $\overline{q}$ and $\overline{v}$. The difference $\overline{q\,v}/g - \overline{q}\,\overline{v}/g$, i.e., the eddy transport term, is not shown by a line. The dashed line lets one see how large the error can become when only data for surface and the so called standard levels are used. The units on the abscissa are $\dfrac{kg\ H_2O}{kg\ air}\ \dfrac{m/sec}{m/sec^2} \times 10^4$.

air rushing down the escarpment as katabatic wind and moving out to the sea often has a vertical extent of less than 300 m.

The possible effect of the second source of error, missing soundings, appears in the data of Table 5.4. Bromwich (1978, 1979) has shown that at most stations surface winds > 20 m/sec are the main cause of unsuccessful attempts, or none at all, to launch a radiosonde. Besides such merciless weather conditions like blowing snow type 39 and visibility near zero, the main obstacle is the extreme turbulence which inhibits the sounding by returning the balloon with instrument vehemently back to the ground. Thus, the results of a series of balloon soundings are biased against very strong surface winds, that means against conditions under which the water vapor transport can be large, in particular where intense cyclones happen to be centered to the west or northwest

of the station. A review of years of results of radiosoundings made at antarctic coastal stations reveals that the Russian radiosonde operators must have
uncanny dexterity, most appropriate equipment, or both, to launch a radiosonde
even under the worst wind conditions. Summer is not a difficult time for anybody, but there are stormy winter months with not one of the scheduled ascents
at MIRNY or MOLO missing. Table 5.4 gives an example. All 62 soundings of the
month were successful, so that it is easy to find out how much in this particular month the computed transport value $\bar{Q}$ would be affected by omission of three
soundings launched with a surface wind speed > 24 m/sec.

TABLE 5.4 Integrated meridional transport of water vapor $\bar{Q}$(kg H_2O m^{-1} sec^{-1})
computed for MIRNY, August 1972; procedure:
a) with all soundings, and using detailed profiles between surface and 850 mb;
b) excluding three soundings with surface winds > 24 m/sec, but using the
   detailed profiles;
c) with all soundings, but disregarding any additional information between
   surface and 850 mb.

Proc.	No. of sdgs.	$\bar{Q}_{total}$	=	$\bar{Q}_{mean}$	+	$\bar{Q}_{eddy}$
a)	62	- 6.8		1.0		- 7.8
b)	59	- 2.8		3.4		- 6.2
c)	62	+ 3.4		4.5		- 1.1

The $\bar{Q}$ values of the Table also suggest that their variability (referring to
time and location) must be large, and that for any further interpretation of
the water vapor transport a month is too short a time interval and a combination of the results of neighboring stations is indicated.

Since up to now only the vapor content of the air has been considered,
the question might arise whether the liquid (and the solid) H_2O content of
clouds contributes significantly to the total meridional transport. The answer
is that in clouds in polar regions the mass of H_2O in the liquid phase is
almost always at least one order of magnitude smaller than the mass of H_2O
vapor, and the mass in the solid phase in clouds is even less. Furthermore,
one has to take into account that any thick cloud layer in the troposphere does
not exist through an entire month continuously over a fixed place, much less
still through a year. Therefore, it can be accepted that for a first estimate
of the poleward transport of H_2O the liquid and solid phases might be disregarded (Peixoto, 1973; Bromwich, 1979). Of course, the same cannot equally
apply to the out-transport of blowing and drifting snow, but that is of no concern as long as only the H_2O supply to the continent is considered. Only in the
complete H_2O mass balance equation of Antarctica does the blowing snow effect
have to appear.

5.2.2 Results of ten stations

The second step towards an acceptable estimate of the amount of water vapor available for deposition and accumulation on a large sector of East Antarctica is to examine a summary of transport values computed for each station. A year has been chosen as reference time for reasons explained in the previous Section, and also because most accumulation studies on the plateau present their results in the form of annual sums. Table 5.5 contains the meridional components of the transport vector, combining the results of stations with equal sign of the three $\overline{Q}$ values. Comparing the data in the second column, $\overline{Q}_{mean}$, with the average isobar pattern in Figs. 4.1(a) and (b), one sees that there is a close relationship between the meridional transport and the positions of the two average low pressure centers offshore from East Antarctica at about 25°E and 105°E which reflect the prevailing flow in the lower troposphere above the boundary layer.

TABLE 5.5 Annual averages of the meridional water vapor transport $\overline{Q}$, total, mean, and eddy flux terms in kg H_2O m^{-1} sec^{-1}.
Group A: SANAE 1972, NOVO 1966 and 72, BAUDOUIN 1966,
Group B: SYOWA 1966/67 and 72, MOLO 1972,
Group C: DAVIS, MIRNY, CASEY, all 1972,
Single : MAWSON and VOSTOK, 1972.

Stations	Longitude	Total	Mean	Eddy
Group A	2°W – 24°E	3.7	6.6	-2.9
Group B	40°E – 46°E	-17.1	-12.7	-4.4
MAWSON	63°E	6.8	8.0	-1.2
Group C	78°E – 111°E	- 7.8	- 4.0	-3.8
VOSTOK*	107°E, at 78.5°S	1.2	1.25	-0.05

*assuming saturation with respect to ice.

The large values for $\overline{Q}_{total}$ and $\overline{Q}_{mean}$ of Group B suggest that along the northwest flank of Enderby Land the substitution of a stretch of a parallel for the real coastline leads to an overestimate of the mean meridional component, because the zonal component of the transport vector is negative (from the east), and the air in the lower (moister) layers tends to move parallel to the coast and the mountains of the hinterland. Positive values at MAWSON appear to be possible in view of the station's location on a meridian about half-way between the prevailing positions of the above mentioned cyclonic centers offshore. However, the computed values are probably too large as consequence of a relatively large number of days without soundings. Additional years' data will have to be analyzed to eliminate these and other uncertainties. Still, the small variation of the eddy terms along the coast is noteworthy.

The results as shown in Table 5.5 could be further used to determine how much precipitation has to fall on the large area between coast and the main ridge of the high plateau, when the relatively moist air masses move upslope and poleward. To do that, one would have to take into account the convergence of the meridional flow and to assume that the VOSTOK values represent the average water vapor content of the air deep inland. This way, the water vapor transport values and the small and relatively well known accumulation data for the area south of the, say, 2500 m elevation line could be used to estimate how much precipitation has to fall on the irregular, extremely difficult terrain between the plateau proper and the coast. Of course, soundings of a fair number of years would have to be analyzed for a comparison with multi-annual accumulation data like those of Fig. 5.4.

5.3 CLOUDS AND FOGS

5.3.1 Cloudiness

Of all antarctic stations operating continuously since their installation, the oldest can look back on more than 80 years when this book appears. ORCADAS, however, also has another, less enviable peculiarity of meteorological interest: it is located in one of the most cloudy places on Earth with reliable weather records, not far from the belt of maximum cyclonic activity on the Southern Hemisphere. Table 5.6 shows a short comparison for "summer" (November-February) and winter (April-September).

TABLE 5.6 Mean total cloudiness ($\overline{N}$) and mean low cloudiness ($\overline{N}_L$) at ORCADAS and three stations on the coast of East-Antarctica, in eighths (octas).

	$\overline{N}$		$\overline{N}_L$	
	summer	winter	summer	winter
ORCADAS	7.4	6.6	6.2	6.3 /8
CASEY	5.8	5.2	4.1	4.3 /8
MOLO	5.4	5.5	1.9	2.2 /8
MAWSON	5.2	4.3	2.3	1.9 /8

The $\overline{N}_L$-values can be of some interest for practical purposes. It can be decisive whether a cloud cover consists only of cirrus and cirrostratus ice clouds, or a low ceiling layer of stratus or stratocumulus which in most cases are supercooled water clouds. Very often, however, average values $\overline{N}$ and $\overline{N}_L$ can be misleading. At many places the mean amount of clouds, or of low clouds, turns out to be between 3 and 5 octas, but such a value is the result of averaging a large number of observations of fair and a large number of cloudy skies, while 3, 4, or 5 octas is observed much less frequently. Therefore, somewhat different statistics of relative frequency values for fair and for cloudy skies can give more usable information (Table 5.7).

TABLE 5.7. Cloudiness: Summary of 3- or 6-hourly synoptic observations in the form of relative frequency (%) of fair and cloudy skies. In this Table summer stands for November to February, and winter for April to September. The number of observations aboard the research ship Eltanin when cruising between 60 and 65°S are 368 in summer and 476 in winter.

Station	Lat	years	Fair scale	Fair summer	Fair winter	Cloudy scale	Cloudy summer	Cloudy winter
ELTANIN	60-65	1962-68	0-2/8	3%	5%	6-8/8	90%	85%
FARADAY	65.3	71-80	0-2/8	11	14	6-8/8	82	77
CASEY	66.3	57-63	0-3/10	21	27	8-10/10	70	66
SYOWA	69.0	66-79	0-2.5/10	21	19	7.5-10/10	48	54
HALLETT	72.3	57-63	0-3/10	27	45	8-10/10	59	45
HALLEY	75.5	71-81	0-2/8	18	32	6-8/8	72	56
LIT. AM.	78.3	40/41,56-58	0-3/10	24	38	7-10/10	67	54
BYRD	80	57-66	0-3/10	25	43	8-10/10	61	43
VOSTOK	78.3	63-67	0-2/10	61	56	8-10/10	24	22
PLATEAU	79.3	66-68	0-3/10	48	65	8-10/10	40	24
S-POLE	90	57-66	0-3/10	49	63	8-10/10	18	11

The variation of cloud conditions with latitude is quite pronounced; it is mainly due to the change from a relatively warm water surface to snow-covered ice or land. FARADAY is typical only for the west side of the Antarctic Peninsula where the maritime influence is still considerable. More surprising might be that the cloudiness at BYRD, with 1500 m elevation and almost 1000 km distance from the nearest water surface, is very similar to that of HALLETT and LITTLE AMERICA, differing strongly from the three stations on the high plateau.

Another question of general interest is: What kind of clouds? Table 5.8 gives the answer at least for one station on the coast and one on the high plateau. It must be added that at VOSTOK as well as PLATEAU Station some cumuliform clouds, a few hundred meters above surface, have occasionally been observed during the short summer.

It is not the author's intention to include here a discussion of the beautiful optical phenomena which, as far as ice particles play a major role, probably nowhere present themselves as colorful and manyfold as in the polar regions. On the plateau, people might live in the midst of an ice cloud that differs from cirrus clouds only by its distance from the surface, though not everyone will realize it. In any case, the purity of the air and the proximity of the multiform, reflecting and refracting ice particles contribute to the

TABLE 5.8 Relative frequency (%) of the observations of clouds in different layers of the troposphere, or clear sky. When there were clouds in more than one layer, only the predominating cloud was counted.

a) VOSTOK, 1958-61 and 63-67 (from Averyanov, 1972)

Layer	Clouds	Dec + Jan	Feb + Mar	Apr - Sep	Oct + Nov	Year
high	Ci,Cc,Cs	54	54	49	61	53%
middle	Ac	7	6	4	3	5
	As	2	4	5	4	4
low	Sc	<2	<1	<1	<1	<1
	St	<2	2	<1	1	<1
overall	clear sky	34	34	42	31	38

b) MOLO, 1963-72 (from Averyanov, 1975)

high	Ci,Cc,Cs	27	18	20	25	22%
middle	Ac	35	29	27	32	30
	As	6	9	15	15	12
	Sc	17	25	16	13	17
low	St,Ns	4	12	8	4	7
	Fog	0.5	0	0	< 0.1	0.1
vert.	Cu,Cb	1	0.2	< 0.1	0	0.2
Indetermin.	X	0.2	1.3	2.2	0.5	1.4
overall	clear sky	8	7	12	10	10

brightness of the phenomena. M. Kuhn's (1970, 1978) descriptions and explanations based on observations and pictures taken at PLATEAU Station and SOUTH POLE are most informative.

Notwithstanding, a few optical phenomena must be mentioned, particularly one whose beauty is minimal, but whose potential danger very real: Whiteout. A description was already given in Chapter 1 (p. 1). The main point is the presence of a closed snow cover and above it a cloud deck dense enough to make the intensity of the incoming light independent of the position of the sun. Under such conditions, the light from above and the sides is about equal to that from the snow surface. The entire effect is most unpleasant, and on crevassed terrain most hazardous, for a man walking or skiing on an "invisible" surface. He may stumble at any time and frequently fall over unseen irregularities. It is even worse for a pilot who, when forced to land, cannot estimate his distance from the surface nor see the horizon. Since there will be no whiteout without a dense water cloud, invasions of maritime air, relatively warm and moist, on the east-side of cyclones are potential harbinger. Statistics about the frequency of occurrence of whiteout are hard to find. It appears that some observers give a special appendage to their synoptic weather reports while others use only the lowest class of visibility (<50 m) in the normal coded form. For NORWAY Station and later SANAE, 20 to 30 whiteout days per year will be a good guess according to Burdecki (1970).

Two other optical phenomena are of some practical interest because they can guide explorers or help navigators in coastal regions, though the importance of that kind of support decreases as the use of helicopters grows. Snowblink, that can be seen from considerable distance, is a whitish glare on the underside of clouds, due to the reflection of light from a snow-covered surface, often a snowed-in ice field. Watersky is almost the opposite, i.e., a relatively dark underside of a cloud deck over open water, a lead or polynya. Both phenomena at the same time, in different parts of the sky, are sometimes called a Sky map.

5.3.2 Low clouds over the high plateau

The cloud statistics in Table 5.8 leave no doubt that the appearance of stratus or stratocumulus as predominating clouds at VOSTOK must be a rather rare event. The extremely small amounts of snowfall on the high plateau are in agreement with such cloud observations. Still, a comparison with the ceiling observations at the SOUTH POLE might be interesting.

TABLE 5.9 Relative frequency (% of all observations taken) of two ceiling classes at the SOUTH POLE. All observations of good visibility with only high cirrus clouds or no clouds at all belong to the second class.

Ceiling class	Dec + Jan	Feb + Mar	Apr – Sep	Oct – Nov
300 – 600 m	7	5	< 2	< 3%
> 3000 m	73	80	90	86%

More information about the low-cloud-cover cases can be derived from an evaluation of direct observations and estimates of the disappearance of the radiosonde - balloon in a 10/10 cloud deck, 33 and 39 cases respectively, of the years 1961 and 1964-68. The mean height of the cloud base above the surface is for the 33 observations: $\bar{h}$ = 430 m, σ_h = 67 m, and for the 39 estimates: $\bar{h}$ = 450 m, σ_h = 130 m. That means the low cloud deck coincides with, or appears in, the warmest layer (discussed in Section 2.4).

This warm layer, by far more pronounced in the long winter than the short summer, is maintained or, after an invasion of colder air from central East Antarctica, restored by adiabatic sinking of air from the upper troposphere and by horizontal advection of warmer (and moister) air from lower latitudes. The important contribution of the latter process can be evaluated from a wind diagram that shows the change of wind with height and hence the thermal wind (see Section 3.1) between 650 and 550 mb. At the SOUTH POLE, this corresponds to about 300 to 1400 m above surface, a layer which is, at least in the winter, above the top of the boundary-layer, so that the time-average of the measured wind vectors represents the mean geostrophic wind vector quite well (Schwerdtfeger, 1968, 1970a).

Fig. 5.3 shows the wind averages of 18 winter days with 10/10 low clouds in comparison to the resultant winds for <u>all</u> winter days with soundings, 1957-66. The resulting thermal advection is computed with equation (5.3):

$$A_T = - \overline{\vec{V}} \cdot \vec{\nabla T} \; \frac{°C}{sec} \quad , \text{ where } \quad \vec{\nabla T} = \frac{f}{R \; \ell n \; \frac{p_1}{p_2}} \; \vec{V}_T \times \vec{k} \qquad {}^{*} \qquad (5.3)$$

A_T for the layer 650 to 550 mb <u>with</u> the low cloud deck amounts to +2.5°/3 hours, about eight times as much as the advective warming rate averaged for all winter days. It is evident that warming rates on the order of 2 to 3° per three hours, and implicitly adequate wind speeds and changes of direction cannot persist for many hours. One sounding per day is certainly not enough to get a complete picture, but may suffice to suggest how the low cloud decks above the surface inversion layer of the antarctic plateau come about.

5.3.3 Fogs

On the antarctic waters away from land or ice shelves, the majority of cases of fog (with visibility < 1 km and vertical extent exceeding the height of the bridge of the ship at which the observations are taken) occur when relatively warm air from lower latitudes moves slowly over colder water, or becomes stationary. Under these conditions the air in the surface boundary layer loses sensible heat, so that the relative humidity increases, the saturation temperature is reached and slightly passed, sufficient to produce and maintain a fog. Dissolution or, at least, decrease of intensity of such fog occurs most frequently under the influence of increasing wind speed and correspondingly increasing vertical mixing with unsaturated air. In anticyclonic circulation systems also light divergent flow of the air over the water surface and a related sinking motion of dryer air from above can take part. Table 5.10 gives a bit more information about fog on the open sea. There seems to be little seasonal variation of the frequency of occurrence of this type of fog, excepting perhaps some increase in the summer when and where the frequency of strong

---

* Notation as in Section 3.1.2, p. 51; $\overline{\vec{V}}$ stands for the vector mean of the winds at bottom and top of the layer.

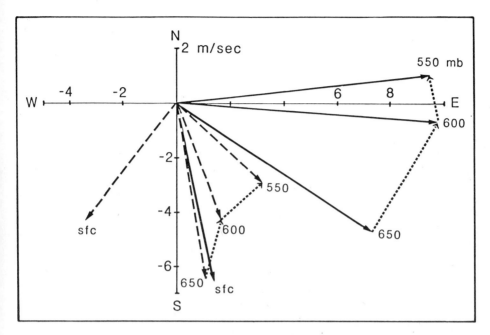

Fig. 5.3. Resultant winds at the SOUTH POLE in 18 cases with 10/10 low clouds (solid lines), and on all days with soundings (dashed lines), both for the winter half years 1961 and 1964-68. The dotted arrows toward the lower pressure numbers give the thermal wind.

winds is smaller. These fogs, often called advection-fogs, can exist equally well in regions with partial ice cover. They also can drift toward land and remain over flat coastal terrain for days. They may be found, of course, in many parts of the world; the stories of the spectacular fogs of the Golden Gate to San Francisco Bay and the dangers of the Grand Banks of Newfoundland are known to every student.

The conclusions one can draw from the data in Table 5.10 are supported by the weather reports of the research ship Eltanin, published in extenso by ESSA (1968, 1970). Altogether, there are 2880 six-hourly synoptic observations made aboard the Eltanin when cruising between 56 and 66°S in the years 1962-68. Of all these, 95 reports (3.3% of the total number) indicate visibility < 1 km (code numbers VV = 90 to 93) and fog (code numbers ww = 42 to 48, most frequent-ly 45). Among these 95 cases of fog, there were 28 with $\Delta T = T_{air} - T_{water} = 0$, 53 with ΔT between 0.1 and 0.5, and 14 with ⩾ 0.6°; none with ΔT < 0. No direct comparison is possible for the mean wind speed, but an average for the 95 cases of < 8 m/sec is certainly a small value for the southern subpolar latitudes. It must be added, however, that this probable relationship -- more wind → less fog

TABLE 5.10   Statistics on fog with visibility < 1 km and duration ≥ 1 hour on the open waters of the South Atlantic south of 60°S (Source: Schwerdtfeger et al., 1959).

a) Occurrence of fog as function of the temperature difference $T_a - T_w$, air aboard ship minus surface water, and of the windspeed. Data from detailed observations at ships of the Armada Argentina.

$T_a - T_w$	<-1	-1.0 to 0.1	0.0 to 0.9	1.0 to 1.9	2.0 to 2.9	>3°	total
Calm	1	-	9	6	8	6	30
1 to 3 m/sec	4	8	18	18	12	11	71
> 3 m/sec	2	1	6	6	2	3	20
All cases	7	9	33	30	22	20	121

b) Duration of fog:

Duration	1 to 3	3 to 6	6 to 9	9 to 12	> 12 hours	total
Number of cases	38	40	18	11	18	125

-- does not hold anymore when the cooling of the initially warm and moist air is intensified by the presence of icebergs.

David Lewis, who solo sailed in spring and summer 1972 from Sydney, Australia to Palmer Island (64°S, 64°W) and 14 months later from there to Cape Town, in the second week of January 1974 came into a large field of icebergs of all sizes, at about 61°S, east-southeast of the South Orkneys. In his book "Ice Bird" (1975) he writes: "... each iceberg I skirted was replaced by another. Fog shut in towards midnight and did not once lift for sixteen hours, and then only to be replaced by snowstorms and whiteout and afterwards by fog again. Ice Bird (his boat) became like a ghost ship. The freezing fog sheathed her rigging with ice; her decks grew ice-glazed, and murderously slippery. Though the conditions had assumed a kind of fantasy nightmare quality, the peril was all too real. Fog, alternating with equally impenetrable whiteout concealed the ever present bergs, while variable head winds and calms prevented us from getting anywhere at all." No further explanation is required. It can only be added that the temperature of the air was about -2°, and that the ice chart of the U.S. Navy Fleet Weather Facility for 3 January 1974 shows 3 to 5 octas ice cover in this region.

An entirely different type of fog, called "sea smoke" or "steam fog", is found over coastal waters, in particular in those regions where strong and cold offshore surface winds prevail and the sea is not, or not yet, ice covered. Among other names sometimes used for the same phenomenon (Huschke, 1959; Saunders, 1964), the terms "cold air advection fog" and "mixing fog" are well chosen because they describe the physical processes which combined initiate and maintain the phenomenon. It is known in wide parts of the world, wherever it is possible that cold air comes to flow over much warmer water. In the typical east antarctic katabatic wind situations, the temperature difference between water surface and advected air will frequently be about 10 to 20°. The dominating processes will then be the continuous heating and the increase of moisture content originating in the lowest layer of air above the warm water, so that convection through, and mixing with, the main body of cold air can develop. Under such conditions condensation sets in, the steam fog forms, and grows with increasing distance from land. The height up to which it reaches will depend on the vertical extent of the cold air and the length of time available. In fact, heights of about 1000 m have occasionally occurred, but in the majority of cases on the Antarctic coasts a few hundred meters might be a realistic estimate, as Figs. 3.10 suggests. Seen from above, the appearance of small, irregular cloud-streets and little cumulus-like bulges is typical, easily discernible from other kinds of fog.

In contrast to open waters, fog over land is an infrequent phenomenon in most parts of the Antarctic. Along the east coast the katabatic winds tend to inhibit fog formation. In the interior, the moisture content of the air is often too small to admit more than a moderate decrease of visibility. The situation changes drastically only when combustion processes, either in aircraft engines or in heating systems of a station, supply sizeable quantities of H_2O and a wide spectrum of nuclei. As far as natural fogs are concerned, the summer months are better disposed for fog than the rest of the year in both regions.

When the temperature in the surface boundary-layer is low enough, the fog observed in the coastal regions can be ice fog, at least in the cold season. Farther inland, ice fog occurs mostly when there is advection of maritime air. For instance, BYRD Station experiences periods of ice fog due to upslope flow from the Ross Sea and Ross Ice Shelf into West-Antarctica. There is no sharp threshold temperature from which on ice particles have to be expected, because the type of available nuclei and the degree of supersaturation are important too. At temperatures below -20° (Mitchell, 1972), according to other sources (Hare and Hay, 1974) below -30°, ice fog has been observed almost exclusively, and below -40° there is no doubt that the existence of super-cooled fog- or cloud-droplets is most improbable for reasons based on the laws of thermodynamics.

5.3.4 Visibility

Determination of visibility at antarctic stations suffers under two difficulties, the absence of landmarks at appropriate distance from the weather station, and the lack of natural light when neither sun nor moon stand above the horizon. These difficulties are insignificant only for poor visibility caused by blowing snow, dense snowfall or fog, because every major station has various constructions scattered over 0.5 to 1 km². Therefore, only two classes of visibility shall briefly be discussed which are of major practical interest and probably also less prone to errors than the others. Relative frequency values are listed in Table 5.11 for three stations with different environment, on the plateau where the moisture content of the air is small, on the edge of the Filchner Ice Shelf where snowfall and blowing snow are much more frequent, and at a place where katabatic winds are less dominant than in other parts of the east antarctic coast. The original statistics published by the U.S. Weather Bureau and ESSA use a scale in eighths of statute miles; this explains the appearance of a 1200 m distance instead of the internationally preferred 1000 m. Such a difference can hardly be important for an estimated quantity like visibility. The advantage of the statistics used for Table 5.11 is that they refer to the total number of three-hourly observations. For many other stations only the average numbers of days with fog or blowing snow or visibility < 1000 m have been published without any discrimination between days with 1 hour and days with 24 hours of fog, etc.

TABLE 5.11 Relative frequencies (% of all synoptic observations in the respective months) of visibility <200 m and between 200 and 1200 m, at SOUTH POLE 1958-67, (28,508 obs), ELLSWORTH 1957-62, (16,018 obs), WILKES 1957-63, (18,253 obs)

Months	Limits	SOUTH Pole	ELLSWORTH	WILKES
Dec and Jan	< 200 m	0.4%	0.8%	0.3%
	200-1200 m	3.4	2.0	1.4
Feb and Mar	< 200 m	1.7	2.9	1.9
	200-1200 m	10.3	8.7	2.1
Apr to Sep	< 200 m	2.5	8.4	9.9
	200-1200 m	7.2	13.9	3.7
Oct and Nov	< 200 m	1.8	2.9	5.2
	200-1200 m	8.4	8.7	2.8

5.4 PRECIPITATION

It is well known that nowhere on Earth can the measuring of precipitation by means of a rain gage or a slightly more sophisticated collecting device be considered a precise procedure. That is so not only because the receiving area of a pluviometer is generally assumed to represent an area 10^{10} times as large, at least. The specific location of the instrument, the different types of precipitation, and the wind play a major role too.

The latter two factors combined are decisive for the polar regions. G.C. Simpson (1919) wrote: "Everyone who has taken meteorological observations in the Antarctic has been faced with the difficulty of finding some way to record the amount of snowfall. Up to the present this problem has not been solved".

There is not much progress to report since. In fact, it has sometimes been observed that a gage was empty after an intensive snowfall and strong winds, whereas plenty wind-driven snow had entered the collector on a day without any real precipitation. Particularly at the stormy places along the coast of East Antarctica the measurement of precipitation is a thoroughly frustrating undertaking, at some stations eliminated from the routine observation schedule. Nevertheless, on the other side of the continent, on the relatively warm northwest side of the Antarctic Peninsula between 63 and 68° latitude, there are at least two stations in less windy locations. The quality of their precipitation measurements is comparable to that achieved at many midlatitude stations of the Northern Hemisphere in wintertime. That means: not very good, but better than nothing.

5.4.1 Plentiful precipitation on the northwest side of the Antarctic Peninsula

The above mentioned two stations are MELCHIOR and AL.BROWN. The former is located on one of the islands of the Melchior Archipelago, between the two large and high islands Anvers and Brabant; several smaller islands shield the station on its west to north side. About 50 km to the southeast, the SW-to-NE oriented main range of the Antarctic Peninsula rises to 2000 m. Upslope precipitation with winds from the northwest sector feeds the glaciers on the mountains. 65 km south of MELCHIOR on the grounds of the Peninsula itself lies the second station, AL.BROWN, in the spectacular, heavily glaciated mountain environment of Paradise Bay. As that name suggests, this probably is the place with the mildest climate of the entire continent, most attractive for antarctic tourism. Position and some data of these two stations and four others less favored by Nature are given in Table 5.12.

Winds from the northwest quadrant in the lowest 2 km of the troposphere bring relatively warm and moist air masses toward the Antarctic Peninsula. This leads, even without fronts, to upslope precipitation on the windward islands and the close-by mountain range. The efficiency and frequency of synoptic situations favoring that kind of flow is directly related to the strength and frequency of a positive pressure difference (at sea level) between two appropriately located stations like ORCADAS (or stations on the South Shetland Islands) and FARADAY. A detailed study of the precipitation in this region (Schwerdtfeger et al., 1959) was stimulated by the extremely warm year 1956 that brought for MELCHIOR an annual sum of 2320 mm. Could it really be that at a place in the Antarctic the annual precipitation amounted to more than 2 m water equivalent? It turned out that in the ten months February to November 1956 the average geostrophic wind component from northwest, at a right angle to the peninsula's mountain range, was 5 m/sec, which is about twice as much as

TABLE 5.12   Position, record, and average values of precipitation, wind speed and temperature at stations on the northwest side of the Antarctic Peninsula.

Station	Lat	Long	Elev. (m)	Years	Average annual values of		
					precip.	wind speed	temperature
Tte. CAMARA	62.6°	60.0°	22	1953-59	568 mm	6.4 m/sec	X
DECEPCION	63.0	60.7	8	1949-60	496	6.1	-2.7
MELCHIOR	64.3	63.0	8	1947-60	1108	3.4	-3.6
AL.BROWN	64.9	62.9	7	1951-59	836	2.2	-2.8* (19)
FARADAY	65.3	64.3	11	1950-53⎱ 1972-80⎰	444	3.7*(25)	-4.7* (35)
MARG.BAY	68.1	67.1	5	7 years from 1946-57	360	4.9	-6.6* (15)

*)these values are from longer series,
number of years in parenthesis.

the multiannual average. The temperature at MELCHIOR was 2.5° above the 12 year mean value. The next question referred to the duration of precipitation; was it compatible with the reported amounts? Fortunately, the synoptic observations at MELCHIOR in 1956 had been carried out with old-fashioned care, and included detailed annotations about time of the precipitation periods' beginning and end. Table 5.13 summarizes what can be derived from these data.

Once one accepts the count of hours with precipitation -- and there is no reason not to do so -- all the computed percentage and rate values appear very reasonable and clearly compatible with the measured sum of precipitation, 2270 mm for the ten months used for Table 5.13, about twice as much as the 14 year average. In this exceptional year, also the mean relative humidity, mean total cloudiness, and still more so the mean amount of low clouds were considerably greater than in other years. Also at AL.BROWN, 1956 brought above average annual precipitation, 972 mm, though the positive deviation was much smaller than at MELCHIOR. It appears quite possible that Paradise Bay is overprotected by the two major islands in this region, Anvers Island (2800 m) to the west and Brabant Island (1900 m) to the north.

5.4.2 Ice crystal precipitation on the plateau

In sunlight or in the beam of a searchlight, even when there is not a single cloud in the sky above a place on the Antarctic Plateau, more often than not one can see a light reflection from small ice particles that drift and slowly fall through the lowest few hundred meters of the atmosphere. There is

TABLE 5.13. Duration and time rate of precipitation on the NW-side of the Antarctic Peninsula in an extremely warm and wet year. Data of MELCHIOR, February to November 1956.

Type of precip.	code ww	hours	percentage of all hours
Drizzle	50-58	357	5%
Rain	60-68	163	2%
Snowfall	70-77	1717	24%
All precip.	50-77	2237	31%
All hours	---	7296	100%
Continuous moderate or strong rain	63 and 65	61	< 1%
Continuous moderate or strong snowfall	73 and 75	673	> 9%
Both		734	10%

Overall rate of precip.	1.0 mm/hour
rate of continuous moderate or strong precip.	2.7 mm/hour
rate of the other ww cases	0.2 mm/hour

nothing new about this phenomenon. Any friend of Nature might have seen it occasionally at sunrise on a clear, cold winter day in continental regions of Northern Hemisphere middle latitudes, and explorers of the Arctic have described it, many years ago. Only on the Antarctic Plateau, however, the clear sky ice-crystal precipitation is sufficiently frequent and persistent to contribute substantially to the total accumulation. As far as frequency statistics are concerned, apparently much depends on the attention of the observers. The alert meteorologists at PLATEAU Station reported 316 "days with" ice crystal precipitation per year, based on their data for 1967 and 1968. Rusin (1964) had related only 159 days with 'precipitation not from clouds' at KOMSOMOLSKAYA (3420 m) in 1957, but 286 such days at SOVIETSKAYA (3570 m) in March to December of that year. The most comprehensive statistics have been published for VOSTOK by Averyanov (1972) and for PLATEAU Station by Radok and Lile (1977). Table 5.14 summarizes the results of nine years of observations at VOSTOK.

Some remarks on a few terms used in Table 5.14 and in the later text might be in place. The literal translation of the heading of the second line in parts a) and b) of the Table (table 45 in Averyanov, 1972) is "ice needles", and says nothing about the absence of clouds. Still, the latter point is answered qualitatively by Table 5.7. The term "ice needles" is often, and in various languages, used instead of the more general word "ice crystals", not necessarily implying that the crystals have the shape of needles. The terms "rime", "glace", and "hoarfrost" are to be understood as follows. When an air mass of maritime origin moves over the ice shelves or large ice fields it will be cooled from below, to reach and eventually slightly exceed saturation in the

TABLE 5.14. Average numbers of a) days per month, and b) duration in hours per day counted in a), for three different processes of precipitation or deposition observed at VOSTOK. Data for the nine years 1958-61 and 1963-67. The term "hoarfrost" is explained below.

Season →	Dec + Jan	Feb + Mar	Apr to Sep	Oct + Nov	Year
observed	a) Average number per month, and annual sum, of days with...				
Snowfall	4	2	2	2	28
Ice Needles	.19	19	21	22	247
Hoarfrost	17	15	21	19	225
	b) Average duration in hours per day counted in a)				
Snowfall		5.5	6	6	---
Ice Needles		11	14	12.5	---
Hoarfrost		15	19	17	---

surface boundary layer. Under such conditions supercooled water droplets form, the visibility decreases, and often fog develops. Some droplets inevitably come in contact with the cold snow or ice surface, and freeze to become a milky, granular deposit. This process is called accretion, the product rime. Small droplet size, a high degree of supercooling, slow accretion, and rapid dissipation of the latent heat of fusion favor the formation of rime. Opposite conditions rather produce glaze, a clear, smooth coating of ice, denser and harder than rime. Finally, when the temperature of the surface or of any object in the way of the moving air is below the frostpoint temperature, direct deposition (change of phase of H_2O from gaseous to solid) occurs. The product, hoarfrost, also "surface hoar" when appearing on a snow- or rough ice-cover, is more feathery and less dense than rime.

Size and shape of the ice crystals under a cloudless sky are the next points of interest. According to M. Kuhn's observations at PLATEAU Station, most of the crystal fall is composed of terminal columnar prisms, often called "bullets". The long axis of these crystals was between 500 and 1000 micron, the diameter about one-third to one-fifth of their length (Kuhn, 1977). Only on cold winter days he also observed a type of delicate prisms which he called "whisker crystals. Their size was estimated at 0.1 mm length and about 0.01 mm diameter. They usually escaped the notice of the observer until they formed a thick, fluffy cover on the ground, on wires and instruments. This cover was very coherent and was sometimes picked up by the wind in larger pieces which would form balls of up to 5 cm diameter rolling along on the snow surface. The density of such balls and of the cover on wires was repeatedly measured to be on the order of 0.01 gram/cm³". It appears these crystals form in the lower part of the inversion layer and possibly continue to grow when on the ground; they "were not observed in the upper half of the 32 m high micrometeorological tower". (The maintenance of the temperature- and wind-sensors at various levels of this tower required one of the observers to climb up even on the coldest days.)

Ice crystal replicas made at the SOUTH POLE, March to September 1975, during clear sky precipitation brought a variety of configurations (Smiley et al., 1980). The larger ones observed were plates, prisms, bullets, and clusters, with sizes mostly between 80 and 225 micron.

It is obvious that ice crystals, growing through molecular diffusion of water vapor (very small Reynolds number), cannot achieve instantaneously such sizes as mentioned above. On the other hand, once they have reached a certain mass, their speed of fall is by no means negligible (Miller and Schwerdtfeger, 1972). For instance, measurements by Yagi (1970) indicate that ice prisms 100 micron in length, developing in water droplet fog at temperatures of -20°, fall with a speed on the order of 10 cm/sec (360 m/hour). Consequently, the ice crystals in the shape of bullets, needles, or plates, seen by an observer at surface, cannot have formed in the lower part of the inversion layer; there would not be enough time for growth. Furthermore, at very low temperatures like -50 or -60° the moisture content of air is extremely small, so that the growth of crystals can only be very slow. Hence, it is much more likely that the crystals form and grow in the "warmest" layer available. Formulating considerations of this kind, Miller (1974) computed the growth time of bullet crystals for various combinations of temperature and supersaturation with respect to ice. Table 5.15 gives some results.

The presence of a warmest layer with temperatures between -35 and -45° and a thickness of 1000, at most 2000 meters during nine to ten months of the year is asserted by the upper air soundings. A state of supersaturation (ice) in this layer is produced, and maintained over long time intervals, by advection of relatively warm, maritime air and by radiative cooling along its way over the Antarctic Plateau, as discussed in Section 5.3.2. In fact, from the winds and thermal winds above the inversion layer at the SOUTH POLE one can compute an average thermal advection between +2 and +3° per day, in individual cases much more (Fig. 5.3). The combination of warm advection and radiative cooling leads inevitably to an increase of relative humidity and eventually supersaturation (ice).

This explanation of the not-from-clouds ice crystal precipitation was confirmed, within reasonable limits, a few years later by the work of researchers at the SOUTH POLE (Smiley et al., 1975, 1976 and 1977; Smiley, 1980; Smiley and Morely, 1981; Hogan et al., 1982). They used modern vertical lidar techniques to determine at what height in the atmosphere supercooled water

TABLE 5.15. Growth time of bullet crystals to a size of 100 × 500 microns as function of temperature and supersaturation. Assumptions: density of air = 1 kg/m^3; molecular diffusivity of water vapor in air = 10^{-5} m^2/sec; a hypothetical spherical crystal growing to the same mass is substituted for the observed bullets.

Temperature			-35	-40	-45	-50°	
		10%	5.5	8.3	17	30	hours
supersaturation over ice	<	20%	2.8	4.2	8.3	15	"
		40%	1.4	2.1	4.2	7.7	"
at saturation (water) which implies			1.4	1.7	3.1	5.0	"
supersat. (ice)			41%	47%	54%	61%	

droplet layers appear or ice crystals develop, and what types, sizes, and concentrations can be expected. The detection limit of the lidar instrumentation was 0.1 crystal per liter of air for a prism of 400 micron length at 1000 m above surface. Various other crystal types have already been named above; it appears that 1 mm maximal extent is about the upper limit for these crystals as observed at the SOUTH POLE. Detailed analysis of the measurements made on six clear sky precipitation days between May and October gave an average diameter of plates and average maximal extent of the other formations from 100 to 200 micron, and concentrations between 6 and 300 crystals per liter. Supersaturation (ice) existed in layers of several hundred meters thickness, in and above the inversion. Multiannual monthly averages of relative humidity in the surface layer of VOSTOK, indicating up to 20% supersaturation (ice) in the non-summer months, were recently published (Zavialova, 1980). Nevertheless, exact measurements of the H$_2$O content of the real atmosphere at various heights by means of modern instrumentation capable to work at the temperatures of the Antarctic Plateau, are still needed.

Large supersaturation (ice) can be expected in and above the boundary layer over the high plateau mostly in the nine or ten "winter" months. In the short summer, the surface inversion is weak or disappears periodically in the afternoon, and the warm air advection above the inversion is less efficient than in the rest of the year. Notwithstanding, every summer has a fair number of days with not-from-clouds ice crystal fall that must originate several hundred meters above the surface. A study of wind and weather in the five summer seasons 1971/72 to 75/76 at the SOUTH POLE (Lax and Schwerdtfeger, 1976) showed that in this part of the year the occurrence of the ice crystal fall depends on the direction of the flow of air in the lowest 1000 m of the atmosphere. That can be ascertained by means of separating the information for all days into a group with and a group without precipitation, and subdividing each group into two classes according to the direction of the vector mean wind at 650 and 600 mb, about 400 and 900 m above surface. This latter subdivision was determined by the topography of the plateau: Class SE = days with winds from the half-circle between 45°E and 135°W, including most of East Antarctica and the Ross Sea sector, that is, winds from terrain several hundred meters higher than SOUTH POLE; class NW = days with winds from the opposite half-circle, including the Weddell Sea sector, that is, from lower terrain. In the two months December and January, real snowfall is a rare event on the high plateau, and on cloudy days one observer might have called intense ice crystal fall what another observer would have taken for light snowfall. That is the reason why, for the statistics in Table 5.16, days with any kind of precipitation were summarily counted for the first group, "days without" for the second.

TABLE 5.16. Frequency of days with ice crystal precipitation at the SOUTH POLE in December and January, depending on the provenance of the air in the 650 to 600 mb layer. Data for 1971/72 to 1975/76. Number of all days with wind soundings = 280.

Group	Class SE	Class NW	Total
Days with	29	150	179
Days without	53	48	101
Sum	82	198	280

The main point is that the movement of air masses with winds of class SE has a downward component in the last 300 or more kilometers, and hence there is adiabatic warming and less supersaturation or none. In contrast, air masses moving and slightly rising toward the Pole from the opposite sector undergo adiabatic cooling and increase of relative humidity, which leads to supersaturation (ice) though not necessarily supersaturation (water); the latter means the sky could remain without low clouds. Naturally, a look at a topographic map of Antarctica might suggest making a subdivision into more than two classes. That, however, could not bring much improvement because what really is needed are trajectories of the air masses, and trajectories can hardly be drawn with data of a single station, the nearest neighbor being 1290 km distant. In any case, the numbers in Table 5.16 leave no doubt that winds from the NW sector favor the occurrence of ice crystal precipitation in the two summer months. This was also confirmed by studies of Smiley (1980).

In this context it should be mentioned that an analysis of nuclei in ice crystals falling at the SOUTH POLE in December 1974 (Ohtake, 1976) indicated the presence of kaolin particles and clay minerals (either from volcanoes or from the Australian desert) on some days, clay minerals and sea salt on others.

The concentration of ice crystals was on the order of 100 crystals per liter. Furthermore, the results of a research project at the SOUTH POLE about type and origin of the aerosol (small particulate material floating in the atmosphere), recently published by Hogan et al. (1982), are of interest. Since 1976 the researchers have also measured the moisture content of the air and determined its dependence on the wind direction all year round, with a much subtler scale of directions than the two class scale described above. Hogan et al. inferred that "the warmest, wettest, most particle-laden air arrives with the relatively infrequent winds from the northwest quadrant". Combining this statement with the result for the summer months alone and with the observation of the high frequency of the occurrence of not-from-cloud precipitation in the non-summer months, one might conclude that in the cold season the conditions for ice crystal formation, growth, and fall are favorable enough to overcome the topographic effect.

5.4.3 Precipitation, deposition, and sublimation on the high plateau

The seemingly simple question "What fraction of the total annual precipitation is contributed by ice crystal fall, and what by (ordinary) snowfall?" turns out to be a rather complicated one. Miller (1974), referring to PLATEAU Station and assuming that favorable conditions persist for 300 days per year, computed a mass increase on the surface due to ice crystal fall on the order of 1 gram/cm^2 per year. That would be about one-third of the total accumulation on the driest and highest part of the plateau. Since the saturation vapor pressure is not a linear function of temperature, -- de$_s$/dT and d^{2}e$_s$/dT^2 are positive -- short periods of strong warm air advection produce disproportionately more ice crystals than computed for average conditions. Therefore, it appears well possible that "one-third of the total" is an underestimate. This is supported by a study of Radok and Lile (1977). They evaluated the diary of experienced observer Robert Dingle who had kept detailed notes on time and type of precipitation at PLATEAU Station for the twelve months February 1967 to January 1968. The amazing result of the investigation is that ice crystal fall contributed 87%, ordinary snowfall only 13% of the total annual precipitation. For a direct comparison with another place on the high plateau, the proportion of time occupied by the two types of precipitation is available. In the nine years used in Table 5.14, ice crystal precipitation occurred at VOSTOK in 36% of the total time, snowfall in 2%. For one year at PLATEAU, it was 53% and 2%. That is a surprisingly good agreement, considering the utter independence of one set of detailed observations from the other, with respect to observers, location, and duration. It strongly suggests that the ice crystal fall really is the major contributor to the total precipitation on the high plateau of East Antarctica, probably at elevations beyond 3000 m.

The next questions may be how much the deposition of hoarfrost on the surface can add to the total income of H_2O, and how much can be subtracted by sublimation. Measurements must be difficult because of the little weight, delicate structure, and fragility of the crystal composites. Any measurements would have to be extended over a long time interval, but even the weakest kind of drifting snow would annul them. Nevertheless, Liljequist (1957) and Loewe (1962) have shown that with some knowledge of the vertical temperature profiles, combined with moderate assumptions regarding the little available moisture, realistic computations can be made. In the lower part of the inversion (the surface boundary layer) there is very little, if any, horizontal advection of moisture. Hence, enduring deposition requires a continuous supply of water vapor from the warmer and moister layers in the upper part of the inversion and above it. This implies that there must be moderate upper limits to the productivity of the deposition process. Loewe (1962) concluded there should exist a downward transport of water vapor < 0.3 gram/cm^2 per year, available for hoarfrost formation.

Closely related is the inverse process, sublimation. Undoubtedly, during 220 to 300 days of every year, depending on latitude and elevation, there is no chance whatever for sublimation from the snow or ice cover of the interior because the presence of a strong temperature inversion in the lowest hundred meters assures that the specific humidity increases with height. In the short summer, however, that is not necessarily so, and there must be a significant difference between the highest latitudes (> 85°S, approx.) and the remainder of the plateau. At 90°S, the periodic diurnal surface temperature variation is zero, because the change of the height angle of the Sun during 24 hours is minimal and not periodic. At PLATEAU Station, at 79.2°S one of the nearest neighbors of SOUTH POLE, the height angle on solstice day varies between 13 and 34°, and the range of the periodic temperature oscillation is on the order of 10° (Fig. 2.6, p. 32). For VOSTOK (78.5°S) the mean temperatures (1958-61, 63-69) are available only for 01, 07, 13, and 19 hs local time (Averyanov, 1972). The mean temperature difference $T_{13} - T_{01}$ amounts to:

Oct and Nov	Dec to Feb	Mar and Sep	Apr	May to Aug
9.0	7.4	4.3	1.1	0.2° .

It must therefore be expected, and can be seen in Fig. 2.6, that in these latitudes an inversion does not exist in the lowest ten meters during several hours of spring and summer days, and hence probably there is no supersaturation at that time.

Loewe (1962) inferred that there is deposition on the Antarctic Plateau during the major part of the year, contributing about as much to the annual increase of the snow cover as sublimation in the short summer takes away. This has generally been accepted in the past two decades as average for the plateau

(Lettau, 1977; Carroll, 1982). To the author's knowledge, it has not been confirmed nor rejected by field measurements. Still, if that holds for the highest latitudes, it might not be realistic at places 1000 to 2000 km distant from the Pole, and should also depend on the prevailing surface wind speed. This notion is supported by recently published results of simple but apparently dependable measurements by Fujii and Kusunoki (1982) who worked in 1977 at MIZUHO Station (70.7°S, 44.3°E, 2230 m, 3/1000 slope of terrain), a very windy place (Table 3.8) where the surface is more ice than snow, and the Sun remains below the horizon for no more than ten weeks, reaching a height angle of 43° at the summer solstice. Consequently, several components of the energy budget and their seasonal variations must differ considerably from those for the SOUTH Pole. One kind of the instruments used at MIZUHO was a set of mini-evaporation pans, small, shallow dishes of thin glass, filled with ice, put into the glazed uppermost surface-layer so that they stand level with the surroundings. The other kind was a mini-stake farm. For the summer time 24 November to 11 January the average loss of H_2O mass by sublimation was found to be 62 milligram/cm^2 per day, in December alone 67, so that approximately 2 gram/cm^2 per summer month appears a reasonable estimate. The efficiency of the deposition process in the winter half-year turns out much smaller, as one might expect for a place with strong inversion winds and low moisture content of the air. For May to July, about +0.2 gram/cm^2 per month has been obtained. For the annual balance a good estimate cannot be given because the two series of measurements are incomplete, but it appears quite acceptable to guess that the net loss amounts to several grams/cm^2 per year.

In summary, it might be concluded that in the wider surroundings of the South Pole, latitude >85°S approx., deposition and sublimation combined contribute but little to the annual H_2O mass budget, and that sublimation becomes the stronger of the two terms at lower latitudes. The latter should not necessarily apply to the ice shelves, particularly not to those parts where the proximity of open water favors the occurrence of fog and the formation of rime (Linkletter and Warburton, 1976).

5.5 ACCUMULATION

As explained in the previous Section, there is not much hope ever to get precise series of precipitation measurements from the major parts of the Antarctic. Considering the practical consequences and also the theoretical interests, it appears justified to ask, "What is really most needed?". There are no agricultural activities, no hydroelectric powerplants, and not much else requesting daily or weekly precipitation data. Furthermore, the shorter the time interval, the greater is the uncertainty: on windy days very little fresh fallen snow remains where it came down. In contrast, reliable information is required about the mass of snow and ice that accumulates every year. It is an absolutely necessary input for the continuous surveillance of the mass balance of Antarctica's ice which affects the stability and motion of the great inland ice sheets, the temperature of the southern ocean with potential influence on the climate of wider areas still, the level of the world's oceans, and so on. There is more discussion of such problems in Chapter 6.

Precisely, the word "accumulation" embraces all processes contributing to the increase of the mass of solid H_2O on ice shelves, sheets, or glaciers. That would include, besides the three processes named in Table 5.14, snow transported by the surface wind toward a reference place, and rainfall, a rare but not impossible event as far south as low elevation terrain reaches (see Section 4.2.5). Correspondingly, "ablation" has the opposite meaning, hence referring to sublimation, wind transport of snow away from the place, and runoff of meltwater; the latter can happen only in the marginal regions of the continent. Whenever one determines how much the mass of solid H_2O has changed since the previous measurement, one really deals with the balance of accumulation minus ablation. Nevertheless, in the polar literature, including these lines, the terms accumulation and ablation are generally used in the sense of positive and negative mass balance.

5.5.1 Measurements

A great advantage of measurements of accumulation in comparison to some other geophysical variables is that several devices and procedures have been and continue to be used, each of them quite different and independent from the others. A detailed discussion of methods and their problems was given by C. Bull (1971). Only a short review will follow here.

The most straightforward method is to measure the increase of the snow cover by means of a large field of stakes, bamboo or aluminum, combined with the determination of the density of the new layer. This procedure is used at all active stations in the interior, mostly in the form of 1 km^2 "stake farms", 100 stakes at 100 m distance from each other, in order to diminish the effect of chance.

The second method pursues the interpretation of the annual horizontal layers of snow of downward increasing age (and density), visible on the walls of snow pits, and on the large snow- and ice-cores produced in recent years at several places. The identification of the individual annual layers, based on the different appearance of the summer and winter crystallization, becomes more difficult and prone to errors on the higher part of the plateau where the total annual accumulation amounts to little and even the summer temperatures at the surface remain far below the melting point. The snow-pit procedure, called snow-stratigraphy, has been applied in nearly all traverses of the continent. Large amounts of data have thus been obtained, mostly since 1957. The results are not always compatible with the ideal of a smooth field of isolines, or with measurements in the same region a few years earlier, but they certainly bring valuable information, often acquired by hard work at inhospitable places. At PLATEAU Station in the summer 1966/67, for instance, there were, besides a stake farm, 114 shallow pits, three down to 1 m, one to 2 m, and one to 10 m, the latter one reaching back to the year 1840, approximately (Koerner, 1971).

The third method, or group of various methods, is based on the physics of radioactive isotopes. A detailed description would be too far from the main topics of this book. Only as examples, two different radiometric methods (Picciotto et al., 1968 and 1971) shall be mentioned. i: Measurements of the vertical distribution of fission products (Maenhout et al., 1979). Use is made of a reference level maintained in the snow by the remnants of the French thermonuclear bomb test in 1954. This way the average accumulation since 1955 can be found unequivocally. Ten years later, when such measurements were carried out in the NNE sector from the S-POLE to about 76°S, the reference level was between 1 and 2 meters below the surface. ii: Measurements of the decay of lead 210, a natural radioactive atom; the approach is similar to the well-known radiocarbon method, but is applicable to much shorter time intervals.

A fourth group of accumulation measurements is based on the presence and varying (with depth) concentration of micro-size foreign particles, deposited on Antarctica a few months after major volcanic eruptions like that of Mt. Agung in 1963.

TABLE 5.17. Examples of annual mean snow-layer density as function of depth. Data of Liljequist (1956) and Dalrymple et al. (1966).

Place	Lat.	Elev.	0	1	2	3	4	m
MAUDHEIM ,	71.1°S,	38 m	0.38	0.41	0.43	0.49	0.53	gram/cm^3
SOUTH POLE		2835 m	0.35	0.38	0.38	0.40	0.42	gram/cm^3

Density increases significantly with depth, is greater for originally wet snow at lower latitude or elevation, increases when strong surface winds can "pack" originally loose snow. In the short and relatively quiet summer 1966/67 at PLATEAU Station, density values as low as .29 gram/cm^3 have been determined for the uppermost layer. However, "in the same summer an exceptionally hard layer with a hardness of more than 60 kg/cm^2 and a density of 0.5 gram/cm^3 covered about 5% of the surface" (Koerner, 1971). This explains why many shallow pits and extended stake farms are needed to produce accumulation values representative for a large area.

The accumulation units used by almost all researchers in the Antarctic are gram cm^{-2} $year^{-1}$, which is equivalent to 10 kg m^{-2} $year^{-1}$ = 10 liter water m^{-2} $year^{-1}$ = 1 cm water $year^{-1}$.

5.5.2 Results

Fig. 5.4 shows the positive H_2O balance of Antarctica's interior. In major parts of the continent, the isolines have been taken over from earlier work, in particular the most valuable maps of Giovinetto (1964) and Bull (1971). Significant modifications have been made in the Ross Ice Shelf area where the recent work of the Ross Ice Shelf Project has brought much new insight (Clausen et al., 1979), and in the region dominated by DOME C (p. 61 and Table 3.10). In the latter, new results (Petit et al., 1982) indicate that the area with less than 5 gram cm year has to be extended northward by about 500 km. Regarding the accumulation at the Pole itself, a significant change of the value up to now accepted, and used in Fig. 5.4, has probably to be made. This change is based on a thorough analysis of the deuterium (or oxygen 18) content in about 900 firn samples which represent the snow accumulated near the Pole since 1887 (Jouzel et al., 1983). The average annual accumulation for the 92 years 1887 – 1978 turns out to be 9.2 gram/cm^2, for the first 44 years 10.0, the following 48 years 8.5.

Over most of the coastal regions including the escarpments up to approximately 2000 m elevation it appears uncertain whether the available values are really representative for the wider surroundings along the paths of the many traverses, or indicate mainly local features. Therefore, no isolines have been drawn beyond the 20 gram/cm^2 per year line, with the exception of the region between 70 and 110°W near 75°S where a fair number of apparently uniform data is available. Five degrees farther south is one of the regions where numerous investigations on accumulation, wind-drift of snow, and related processes have been carried out, at the "old" BYRD Station (80°S, 120°W, 1510 m) 1957-62, at new Byrd (10 km apart and 20 m higher), and the wider surroundings (Budd et al., 1966; Benson, 1971; Whillans, 1975; more references in these publications). Here the most important result may be the finding how the differences in accumulation over short distances are related to the minor ridges and depressions and the surface winds. The topography itself is migrating so that the annual accumulation could vary even if the annual snowfall were constant.

For the major part of the Antarctic Peninsula again a big question mark has to be made as it was done in earlier maps. Average annual precipitation sums have been plotted for a few stations along the western side of the

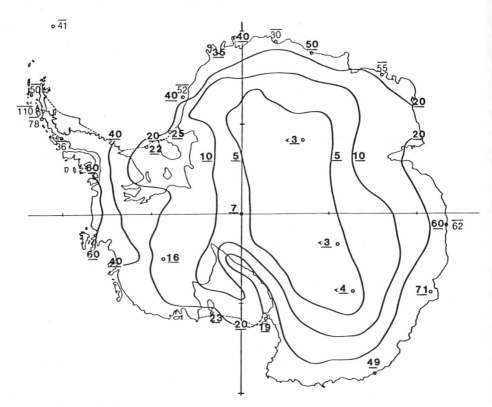

Fig. 5.4. Accumulation (isolines and <u>underlined numbers</u>) in gram cm^{-2} year^{-1}. Average annual precipitation values in the same units, marked by the line <u>above</u> the numbers, have been added for a few stations.

Peninsula (Table 5.12), a poor substitute for comparable accumulation values. Ten year average annual precipitation sums can also be computed from monthly publications for HALLEY, MOLO, and WILKES. Such values have been added in Fig. 5.4, with the reservation that they have not yet been ascertained, which at best could be done in the manner described for MELCHIOR Station in Section 5.4.1.

A new accumulation map for the Ross Ice Shelf region (Fig. 5.5) gives a remarkably detailed picture (114 core positions) of the mass of snow loaded annually on the ice shelf. Indirectly, the map also contains valuable information about vertical motion in the lower troposphere. On the larger scale, it is evident that the isolines of equal annual accumulation (10, 12, 14 gram/cm^2) are nearly parallel to the Transantarctic Mountains from 79°S/160°E to 86°S/160°W, an unsurmountable wall for stable air masses approaching from the east

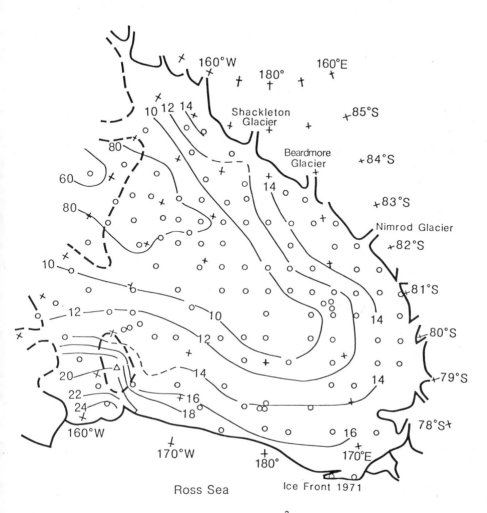

Fig. 5.5. Surface accumulation rate (gram/cm^2 per year) distribution on the Ross Ice Shelf, as measured by total β activity analyses of ten meter firn cores obtained during four R.I.S. Project field seasons between 1973 and 1978. The 114 small circles indicate the position of the cores. The individual results which figure in the original map had to be omitted to keep the repro- duction readable. Near the outlet of the Beardmore Glacier (to the east and north) there are four stations with values > 20 gram/cm^2. All this information is used with the friendly permission of Dr. D. Dansgaard.

or northeast. Over a distance of 200 to 300 km at a right angle to the isolines, the accumulation increases from less than 10 gram/cm^2 in the central part of the ice shelf to more than 15 near the mountains. This indicates the occur- rence of rising motion due to the piling up of the stable air masses (Fig. 3.15, p. 85). Sinking motion must prevail on the eastern side of the Ross Ice Shelf where between 82° and 84°S the annual accumulation is even smaller than at the

SOUTH POLE. All this clearly supports the barrier wind concept (Section 3.3).
On the smaller scale, the comparatively high accumulation values in the northeastern corner of the ice shelf, the region of FRAMHEIM and LITTLE AMERICA, are
obviously related to the presence of the big snow and ice block called
Roosevelt Island (79°S, 162°W) and the advection of moist air masses from the
north quadrant. This phenomenon immediately reminds of Tor Bergeron's analyses
of oreigenic precipitation in the hills of central Sweden, (Bergeron, 1960;
Andersson, 1981), and of Weickmann's (1981) investigation of the snowfall from
shallow, upslope cloud systems along the eastern slopes of the Rocky Mountains
in Colorado.

Chapter 6

SELECTED PROBLEMS OF ANTARCTIC CLIMATOLOGY

Perennial presence of ice and snow can be understood as a major determinant of the climate of Antarctica. Minor variations of the temperature regime in the coastal regions and over the southern ocean are related to the varying abundance of ice. Major changes within limited time, say a few hundred years, appear to be possible only if a sizable part of the ice sheets covering the continent should become unstable. In this context, the word "unstable" implies the possibility of the occurrence of surges, rapid advances of large ice masses, similar but on a much larger scale than observed at various glaciers in the Northern Hemisphere. Such surges can be brought about by "basal lubrication", i.e., melting at the bottom due to an increase of mass by accumulation on top, and by geothermal and frictional heating. A surge of the west antarctic ice mass could eventually lead to a rise of the worldwide sea level by about six meters, a disastrous, though remote, possibility which has received much Cassandra-style publicity in scientific and science-fiction publications. At present, neither the evidence in situ nor the sophistication of models suffice to justify any pessimism. Naturally, surges could also be possible for parts of the much larger east antarctic ice sheet, but specialists consider such an event even less likely than for West Antarctica. Still, it might be mentioned that there is some evidence of a fast rise of sea level by about 20 m, 95,000 years ago, and that an east antarctic surge could have been the culprit (Hollin, 1972; Flohn, 1978).

Considering such far-reaching possibilities, one should not forget that Nature has some built-in brakes, -- feedbacks --, which often prevent unidirectional developments with catastrophic consequences. For the problem of possible antarctic surges it should be taken into account that an increase of tropospheric temperatures would lead to an increase of precipitation and accumulation, hence with much time delay more calving of ice, hence more cooling of the subantarctic waters, hence a decrease of air temperatures, hence less precipitation, and so on. This way, there could be long range rhythms or modest, irregular variations without the necessity of any sudden catastrophic events. Nevertheless, considerations of this kind appear to be interesting enough to make the ice mass budget of the continent, and possible temperature variations on and around it, the first topics to be discussed in the following Sections.

6.1 THE ANTARCTIC ICE MASS BUDGET

This is one of Antarctica's major unsolved problems waiting for further research. Loewe (1956) examined the quantitative contributions of the various

processes which together result in the estimate of the average annual budget
for the entire continent - including the ice shelves: accumulation, discharge
in the form of icebergs, liquid run off due to surface melting near the coast
in summer, sublimation (evaporation) and deposition in the ablation areas, and
bottom melting of ice shelves. The surprising result was a positive value,
meaning a growth of the snow and ice cover of the continent with time, under
the physical conditions of recent years. However, the sum of uncertainties in
the various components turned out to be larger than the computed value itself.
In fact, none of the researchers named in Table 6.1 and the other authors of
eleven independent studies whose results are combined in the Table under the
heading "earlier" has claimed his result to be positive beyond error margin.
That was the situation for twenty odd years since Loewe's first approach. Then,
Kotlyakov et al. (1977) came up with a negative mean annual budget, but added:
"the low accuracy of individual components of the budget does not permit a
categorical conclusion to be made regarding its sign."

Possibly, detailed mass budget studies of smaller drainage areas which
contribute a relatively large part of the total outflux will help to refine the
estimate of the greatest, and least certain, item in the overall budget. Reporting about such an investigation, Budd (1975) writes: "...for the total drainage
basin (Amery) it still appears that the net accumulation may be 50% higher than
the net outflux through the Lambert Glacier." Morgan and Jacka (1979) come to
similar conclusions, based on measurements south of the 2000 m height-line,
southwest of MAWSON. They also quote reports on three other field studies in
parts of East Antarctica which likewise resulted in a positive imbalance, i.e.,
considerably more mass input than output. In summary, the problem is still wide
open.

Table 6.1 shows at the first glance that only the calving of glaciers
and ice shelves can be the component large enough to change the budget interpretation so drastically. That might be quite fortunate because the ever progressing satellite technology should soon be capable of monitoring number, size, and
track of icebergs, from the cradle to the grave, much more efficiently than in
years past. That would be of great importance not only for the ice budget problem, but also for the understanding and forecasting of the subantarctic temperature regime. There is more discussion about this regime in Sections 6.2 and
6.3.

It should be noted that in the upper line of Table 6.1 the net accumulation values are listed. That means the snow removal by the wind is not to be
counted, and only for ablation areas the possible loss by evaporation or sublimation, and in coastal regions in the summer also by the liquid (and not re-
freezing) run off, is a budget component. The size of the total of all ablation

TABLE 6.1. Estimates of the mean annual antarctic ice mass budget.
Units: 10^{12} kg $\approx$ 1 km^3 of water. References: Kotlyakov et al., 1977; Bull,
1971; Giovinetto, 1968; Loewe, 1956 and 1967.

Year of publication Authors	1977 Kotl.	1971 Bull	1968 Giov.	earlier Others
Net accumulation	+1980*	2080	2100	1900 km^3
Run off and evap. in ablation areas	- 10	- 10	0	0
Bottom melting	- 320	- 200	- 200	- 300
calving	-2400	-1450	-1000	-1100
net ice mass budget	- 750	+ 420	+ 900	+ 500

*a 20 km^3 water equivalent of estimated snow removal by wind has been
subtracted from Kotlyakov's accumulation value.

areas is one more problem. There are parts of the coastal belt of land with
furious katabatic winds, some regions near the upper limit of the steep escarpment where continuous strong, unidirectional winds keep the bare ice free of
snow most of the time, and innumerable small ablation areas in the mountains.
Bull (1971) estimates the total area of upper surface ablation probably to
exceed 70,000 km^2; Kotlyakov et al. (1977) say 0.6% of the total surface area
must be considered for the budget, about 80,000 km^2. More than one half of the
latter value has been attributed to the enormous Lambert Glacier that feeds the
Amery Ice Shelf (72°E longitude). Meltwater drainage channels have been found
on the glacier's surface up to about 1000 m above sea level, hundreds of kilometers from the coast (Budd et al., 1967).

Still, all research work done since the IGY confirms that an accurate
estimate of the annual calving from the continent is the key to the "sign of
the budget" question and to various related problems of climatology.

6.2 ICE SUPPLY AND CLIMATE

In February 1823, the whaler James Weddell sailed far south in the sea
that later was given his name, and reached 74°15'S, 34°16'W on the 20th. In his
narration (Weddell, 1827), he wrote:
"This climate appears to be in general much more temperate now than it was
forty years ago, the cause of which may probably be, that immense bodies of ice
were then annually found in the latitude of 50°. This ice, passing to the
northward between the Falkland Islands and South Georgia, would necessarily
lower the temperature of both air and water, and consequently an unfavorable
opinion of the climate was produced."

In 1891, E. Du Faur wrote about "the possible effect of the varying position of floating ice and ice-fields in the Great Southern Ocean on the meteorology of the southern portion of our continent", and concluded "for there is not

the least doubt that in our climatic changes (referring mainly to the occurrence of droughts) we are closely linked with the meteorological conditions of the great polar region so close to our doors" (quoted from Du Faur, 1907; also see Burrows, 1976). Apparently, after the manifold discoveries made in the first half of the 19th century the time for such an idea had come. Of course, the time difference between the coming up of an idea and the test of its validity can be large. In this case, satellite measurements, drifting buoys and other tools, and the general increase of public interest in environmental problems will contribute to a clear answer in the foreseeable future.

When Du Faur wrote about a possible effect of antarctic ice, climate variations in Australia were on his mind. At present it appears the South Atlantic sector should be the first testing ground. Nowhere else the conditions are equally favorable for sizable supply and appropriate transport of ice. Only in the Weddell Sea there is a large area where the pack ice persists through the summer. According to Ackley (1979b), the Weddell Sea contributes more than all other pack ice regions together to the total year-to-year variability of the size of the entire pack ice belt.

6.2.1 Ice in the South Atlantic

In Chapter 3 it has been shown how the so-called barrier winds along the eastern flank of the Antarctic Peninsula originate, and how the alternating occurrence of these winds and the about 20° warmer west winds strongly favor the drift of ice from the northwestern part of the Weddell Sea towards the eastward currents and winds at lower latitudes. This corner of the Weddell Sea is indeed the only region in the realm of the subantarctic seas where there is a kind of "guided tour" available for sea ice and bergs from, say, 70° to 63°S (Schwerdtfeger, 1977, 1979). The striking difference between the ice drifts from the northern part of the Ross Sea westward around Cape Adare, and from the Weddell Sea northeastward, has been described in detail by Ackley (1979b). Furthermore, large masses of moveable ice are almost always present in the northern half of the Weddell Sea, not only drifting sea ice, but also bergs in all sizes coming either from the various ice shelves and glaciers enclosing three sides of this great embayment, or from the Fimbul Ice Shelf and neighboring smaller shelves east-northeast of Cape Norvegia and farther east, not far from 70° latitude. The meteorological observations of MAUDHEIM, NORWAY STATION and SANAE confirm the prevalence of adequate winds (Table 6.2).

The ice supply from the off shore regions to the west of Riiser-Larsen Peninsula (34°E) is probably considerably smaller than that from the Weddell Sea itself. However, there was one interesting change in the route of drifting ice near the 0° meridian, due to the breakaway of Trolltunga: On all geographic

TABLE 6.2 Average annual surface winds (m/sec) at MAUDHEIM, 71.1°S, 10.9°W,
38 m, Mar. 1950 - Jan. 1952; NORWAY ST., 70.5°S, 2.5°W, 56 m, Mar. 1957 - Dec.
1961; SANAE, 70.3°S, 2.4°W, 52 m, Jan. 1963 - Dec. 1967.
n = number of observations; other notations as in Chapter 3, p. 42.
Sources: Hisdal, 1958; Weather Bureau of South Africa, 1966; Burdecki, 1969.

Place	D_R	V_R	V	q	n
MAU	101°	4.6	7.7	.60	5272
NOR	114°	5.2	8.7	.59	14085
SAN	104°	4.5	8.5	.53	12644

maps of Antarctica printed in the 1950s or 60s, one finds near the meridian of
Greenwich a remarkable tongue of ice, 40 to 50 km wide and reaching about 100
km farther north than the main rim of the Fimbul Ice Shelf, doubtlessly a formidable obstacle in the way of a coast-parallel, generally westward current and
wind drift.   In the winter of 1967, Trolltunga went AWOL. As it befits the era
of satellites, Trolltunga was soon rediscovered and observed on its journey and
long stay, 1970-1975, when it was grounded at 77°S near the rim of the Filchner
Ice Shelf. Then Trolltunga came free once more and moved toward north in the
westernmost part of the Weddell Sea along the Larsen Ice Shelf (see Fig. 6.1),
was sighted in August 1977 east of King George Island (62°S), May 1978 at 50°S
north of South Georgia, and possibly in December 1978 in the form of 15 to 20
smaller bergs at about 43°S, 11°E (Vinje, 1979). That is certainly an interesting story by itself, but the main point here to be made is the possibility
that the removal of Trolltunga from its original location could have a significant effect on the transport of ice from the east into the Weddell Sea circulation proper.

        From all this as well as the weekly ice charts of the Naval Polar Ocean-
ography Center it is evident that every year large ice masses are carried into
the southern South Atlantic and the westernmost part of the Indian Ocean, where
they melt. That process requires plenty of energy, which all year round is
available in the warmer ocean water the ice is floating in, and in the summer
months also in the form of solar radiation. The latter can hardly be more than
100 MJ m^{-2} per summer, capable of melting about 30 cm of ice, part of which
would be restored by snowfall in the remainder of the year as long as the ice
has not yet moved too far north. In comparison, the heat supply of the ocean
water is almost inexhaustible because convection in the waters, and different
drift effects of currents at some depth and winds at surface, can always bring
new water close to the ice.

        The magnitude of the cooling of the water due to the melting, and of the
possible modification of the temperature field over the southern South Atlantic,
can be estimated only when there is some information about the ice masses

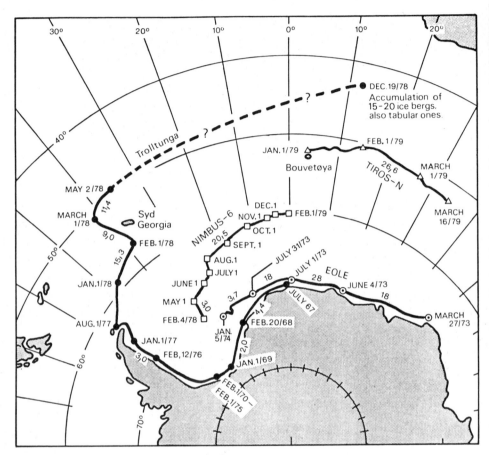

Fig. 6.1. Trajectories of TROLLTUNGA and other icebergs (reproduced with the author's permission, from Vinje, 1979).

advancing northward beyond, say, 64°S. That is still today an inaccurate under-taking. As far as sea ice is concerned, Vinje (1980) estimated the outflow from the Weddell Sea in 1979 to cover an area of roughly 2.5×10^6 km^2. Regarding icebergs, only the largest ones (maximum horizontal extent greater than 15 nau-tical miles) are listed with area estimates in the weekly ice charts (Fleet Weather Facility, later Naval Polar Oceanography Center, 1973-82). For instance, in March 1978 there were four such giants on their way between offshore Larsen Ice Shelf (66.7°S, 58°W) and the sea west of South Georgia (53.7°S, 40°W). The total area of the four, as seen from above, is about 9×10^3 km^2. A satellite picture of a similar monster, a year earlier, is shown in Fig. 6.2. Concerning the total volume of the many ice bodies in size between the shallow (1 to 3 m) sea ice proper and the biggest bergs (thickness about 300 m), the uncertainty

Fig. 6.2. Tabular iceberg, about 65 × 35 km^2, in the northwestern Weddell Sea, an ERTS-satellite picture of end of January, 1977. In the following six months the berg drifted toward NNE from 65 to 62°S, a little more than 2 km per day. No major change in size was observed. Source: NASA, 1977, in AJUS 12(3):3.

is great. Only to continue the line of reasoning, it will be assumed that the total volume of all ice to be melted in an ice-rich year north of the Weddell Sea between 60 and 30°W longitude be $V_i \doteq 4 \times 10^{12}$ m^3.

Such an assumption makes it possible to speculate how large a mass of ocean water, m_w, would be cooled (ΔT_w negative) when a given mass of ice, m_i, is warmed to the melting point (ΔT_i positive) and fused into the liquid state. It is obvious that for this simplified model the following formula holds:

$$\frac{m_w}{m_{ice}} = \frac{c_i \, \overline{\Delta T_i} + L_f}{c_w \, \overline{\Delta T_w}} \quad , \tag{6.1}$$

where c_i and c_w are the specific heats of ice and water, and L_f the heat of fusion. The estimate of the necessary warming of the ice before it can melt remains somewhat vague because ΔT_i is certainly much larger for a big iceberg moving northward than for shallow sea ice. It is here assumed to be a modest + 2° because in most years the contribution of the sea ice will be larger than the contribution of the bergs. A greater value of ΔT_i would still add to the cooling potential of m_i, but this increase would not be large since $L_f >> c_i \overline{\Delta T}_i$. For a simple computation it is postulated that $\overline{\Delta T}_w = -1°$. Under these conditions

$$m_w \doteq 80 \ m_i \quad , \tag{6.2}$$

and introducing the average density of ice and the volume $V_i = 4 \cdot 10^{12} \ m^3$, the volume

$$V_w \doteq 300 \cdot 10^{12} \ m^3 \tag{6.3}$$

That could mean a water body of 300 m thickness and $1000 \times 1000 \ km^2$ area or, more likely, a smaller thickness and a larger area, a smaller thickness and a larger $\overline{\Delta T}_w$, or both.

Without any claim of accuracy or consideration of all details, it appears justified to say that the cooling potential of the ice masses leaving the Weddell Sea is large enough to strongly affect the temperature field of the southern South Atlantic, and much larger than in other sectors of the Antarctic which do not have a natural guidance into the latitudes of the westwind drift.

These two statements can be supported by observed facts.

i: In the zonal band between 54 and 55°S, there are two stations with a long record of observations, USHUAIA at 68.3°W, entirely out of reach for the drift of ice and cold water from the Weddell Sea, and GRYTVIKEN at 36.5°W, precisely in the main stream. The latter is situated on the north side of South Georgia Island, protected by a high WNW to ESE extending mountain range from the direct impact of the cold southerly or southwesterly winds. USHUAIA, on the Beagle Channel of southern Tierra del Fuego, has very little protection, and lies a little more than half a degree latitude farther south than GRYTVIKEN. Notwithstanding, in the 51 year series of annual averages since 1931, GRY is colder than USU by 3.6° ($\sigma_{\Delta T} = 0.4°$), and there was not one year among the fifty-one in which the difference was less than 3.0° or more than 4.9° (Fig. 6.5).

ii: Another way to prove the Weddell Ice effect in the southern part of the South Atlantic is based on data taken from the Climate Atlas of the Southern Hemisphere (Taljaard et al., 1969). With the grid point temperature values (every five degrees latitude and longitude) interpolated from data of land stations and ships, it is easy to show (Fig. 6.3, from Schwerdtfeger, 1979) how the mean annual temperature varies along the parallels 60 and 50°S. The graphs do not require much commentary, but it should be stressed that the values are

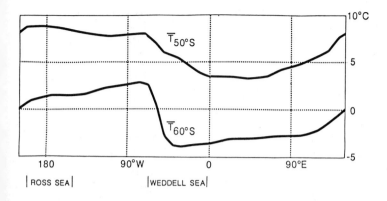

Fig. 6.3. Zonal profile of the mean annual air temperature near sea level
along the parallels 50°S and 60°S.

air temperatures near the surface of the southern ocean. That is necessary
because comparable records of surface-water temperatures do not exist, and is
most informative because the main purpose of this section is to ascertain the
existence of an effect on the atmosphere. In the same latitudinal belt, almost
40° of longitude east of GRY, lies Bouvetoya, 54°26'S and 3°24'E, an inhospitable island of great interest for several branches of geophysics. Only in the
last years some coherent meteorological observations have been obtained for
that place. Vinje (1981) describes wind, pressure and temperature conditions
in 32 months 1977/78 and 1979/80, and finds: "Comparisons show that the present
observations of the surface temperatures are 1 to 3°C lower than previous estimates for this area". This is confirmed by the records of the Argentine station
CORBETA URUGUAY on Thule Island, the southernmost of the South Sandwich Islands
(Tables A 1 and A 3). A complete series of monthly mean values is available for
1979 and 1980. For these two years the mean annual temperature is - 4.8°, only
0.2° higher than ORCADAS (1.3° lat. farther south), and 7.0° lower than
GRYTVIKEN (5.2° lat. farther north).

iii: In this context it is worthwhile to show that the effect can be found
also in the free atmosphere (upper part of Table 6.3). The main point of the
statistics is that at the 500 mb level and at 50°S the lowest temperatures
appear farther east than at surface and 60°S. That makes eminent sense in consideration of the strongly prevailing westerly winds between the two levels and
parallels (Schwerdtfeger, 1977).

TABLE 6.3. Multiannual mean temperature at 500 mb and near the surface, along the parallels 50 and 60°S, based on data of Taljaard et al. (1969).

Longitude	90°W	45°W	0	45°E	90	135	180	135°W
T_{500}, 50°S	-24.3	-25.1	-25.0	-24.9	-25.2	-24.1	-23.9	-24.3°
T_{500} 60°S	-29.1	-30.2	-31.1	-31.6	-31.4	-30.2	-28.7	-28.8
T_{sfc}, 50°S	8.0	5.8	3.4	3.1	4.6	7.0	8.6	8.6
T_{sfc}, 60°S	2.6	-3.4	-3.5	-3.1	-2.7	-1.3	1.2	1.7

6.2.2 Ice supply and temperature anomalies

Whereas nobody doubts that the relatively low temperatures at the Orcadas, South Sandwich, South Georgia, and Bouvetoya Islands are due, at least in part, to the abundant supply of ice from the Weddell Sea into the southern South Atlantic, two other questions still remain unanswered. i: How far northward does a significant cooling effect extend? ii: What can be the reactions of the atmospheric circulation in the subpolar and middle latitudes when there are variations in the annual surface heat budget of a sizeable sector of the southern ocean?

Even if exact values of the yearly amounts of ice available for the cooling of the waters of the southern ocean are not yet known, the satellite pictures of recent years and the weekly ice maps elaborated by the Naval Polar Oceanography Center ascertain that sizeable variations happen from year to year and over longer time-intervals. Under these circumstances research is needed to determine how far downstream or downwind connections with such variations can be found, what feedback effects could be expected, etc. The first step might be to examine the correlations of the series of annual temperatures at various islands or along a coast. Series of data discussed in Section 6.3.1 show that there is a close relationship between temperature and presence of ice near the Orcadas. About 900 km to the northeast lies the station GRYTVIKEN on South Georgia Island, also with a long series of meteorological data, though less homogeneous than ORCADAS. For the 71 years 1911-81, the correlation coefficient of the annual temperatures of the two places is r = 0.62. This, however, appears to be the only significant temperature relationship in the wide region northeast of the Weddell Sea. Much farther to the east, the correlation of the annual temperatures of GRYTVIKEN and MARION Island (46.9°S, 37.9°E), 1949-81, is close to zero, and to the northeast GRYTVIKEN versus GOUGH Island (40.4°S, 9.8°W) r = +0.27 for the 26 years 1956-81 also insignificant. If there are any distinctive relationships, they must be more complicated than simple air temperature correlations.

One possible development has recently been pointed out by Fletcher et al. (1982): the approximately homogeneous cold water masses of the subantarctic spread into lower latitudes at intermediate depths, and their subsequent inter-action with surface layers can help to transfer temperature changes from high to low latitudes. One might add that the surface temperatures of the tropical South Atlantic are low in comparison to those of other tropical oceans. That fits well with the observation that the ice transport from the Weddell Sea toward lower latitudes is greater than from other, comparable parts of the Antarctic Ocean. Whether annual or longer-term variations of the ice conditions in the Atlantic sector are also reflected (with time delay) in the small varia-tions of the surface water temperatures at much lower latitudes, has not yet been investigated, to the author's knowledge.

Until now, the main interest has been directed to the possible conse-quences of the presence of much ice, or little, downstream of a major supply region. It is equally worthwhile to look at the temperature anomalies at coastal stations and the extent of the pack ice belt in the respective sector (Budd, 1975; Jacka, 1981). The latter study suggests that the annual mean tem-perature of a place at the coast is correlated with the latitude of the maximum sea ice extent 10 to 40° longitude farther east than the respective station. This implies that the drifting winter ice reaches sufficiently far northward to be displayed more by the eastward circumpolar current than by the opposite nearer to the coast.

6.3 TEMPERATURE FLUCTUATIONS

And if you have nothing to worry about,
Antarctica's warming might help you throughout.

Of all variables measured routinely at the antarctic meteorological sta-tions, temperature is the most informative for a study of minor variations of the climate in recent years. Precipitation measurements depend far too much on wind conditions and topography; these and other difficulties are mentioned in Section 5.4. The utility of atmospheric pressure values is greater for synoptic meteorology than for climatological studies of minor variations because in the former field one has to deal with changes by tens of millibars, in the latter by small fractions of such amounts. Excepting the year 1979 with the air-dropped buoy experiments (page 10), the scarcity of measurements on the vast area of the southern ocean is a handicap for the full interpretation of the pressure field north of the coastal stations; this is so at any time scale. Another handicap refers to the measurements of atmospheric pressure themselves. The questionable appearance of "pressure jumps" when violent winds suddenly embrace a weather station (Fig. 3.9, p. 66) is discussed on page 71.

6.3.1 Annual temperature series

ORCADAS is the only meteorological station south of 60°S with a long record of observations. It is located on Laurie Island in the South Orkneys group, on a low and narrow strip of land between the two major parts of the island (Fig. 6.4), the western one-third with precipitous peaks up to 900 m, the eastern two-thirds heavily glaciated. To the NW of the station lies the Bahía Uruguay, to the SE the Bahía Scotia. The isthmus itself, between the two embayments, must have appeared to be the only accessible and suitable location on the island; it is about 400 m wide and only 4 m above sea level, which could be considered a dangerous exposure. However, in the eighty years of the station's history, to the author's knowledge, only once (in the 1920's) has the sea overflooded the entire isthmus, including the floor of the one-story building.

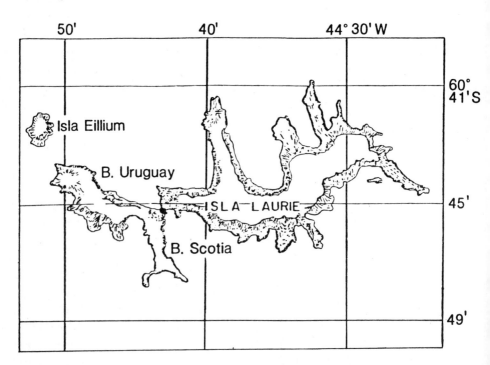

Fig. 6.4. Laurie Island

The ORCADAS station began in March 1903 as winter quarters of the Scottish National Antarctic Expedition 1902-04 (Mossman, 1907; Prohaska, 1951) and was transferred in 1904 to the Argentine Servicio Meteorológico Nacional.

The place of the first refuge and later buildings and the outdoor instrumentation has never been changed. The series of data can be considered as homogeneous throughout. In all the time there was only one interruption, caused by a fire (no victims) in the winter 1975; for a few subsequent months the main series of mean monthly temperature and pressure have been completed by the values of SIGNY Island, a British station in operation since April 1947, located 48 km due west of ORCADAS.

Fig. 6.5 shows that the variability of the annual mean temperatures at ORC is relatively large, the standard deviation $\sigma = 1.1°$ for the series 1904-81. Annual temperatures of island stations in the southern ocean at about 6° lower latitude vary much less, GRYTVIKEN (1905-81) $\sigma = 0.48°$, MACQUARIE (1951-82) $\sigma = 0.38$, USHUAIA (1931-81) $\sigma = 0.44°$. The situation at ORCADAS is different because the presence or absence of sea ice around the South Orkney Islands group determines the strength of its temperature regime. This can be proved by a unique series of observations which was started in 1903 but seems to have fallen in disregard in recent years: the annual duration of an ice cover on Bahía Scotia. The average duration is 180 days, $\sigma = 67$ days, for the series 1903-74 (with four years of uncertain data). Only once has the bahía not opened in summer, remaining closed from 13 May 1928 to 5 November 1929.

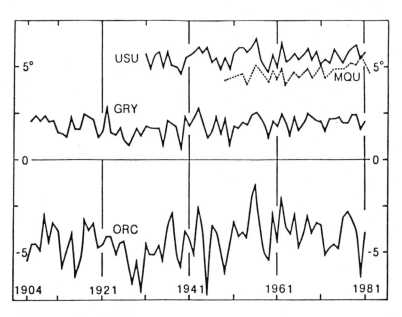

Fig. 6.5. Annual temperatures at ORCADAS (1904-1981), GRYTVIKEN (1905-1981), USHUAYA (1931-1982) and MACQUARIE (1949-1982).

All this is brought up because a solid negative correlation exists between this duration and the annual mean temperature of ORC: $r = -0.80$; short ice duration, warm years; and vice versa. No assertion is here intended whether one or the other of the two variables can be considered the "cause". The long series of the ice data for the atlantic sector of the Antarctic has found much interest, and different interpretations by various researchers (Prohaska, 1951, 1954; Schwerdtfeger, 1959, 1970, 1975; Lamb, 1967; Fletcher, 1969). It can now be stated with certainty that i: the ice conditions observed at ORCADAS are not characteristic for other sectors of the pack ice belt of the southern ocean, and ii: the correlation between ice duration at ORCADAS and T_a at SANAE is positive (for the period 1957-1974, $r = +0.5$), and between $T_{a,ORC}$ and $T_{a,SAN}$ negative![*] The annual temperatures of ORCADAS in Fig. 6.5 (no smoothing applied) indicate a cooling from 1908 to 1930, warming from 1931 to 1956, then again modest cooling for the following 25 years. Such variations do not support the popular hypothesis of the carbon dioxide increase in the atmosphere leading to a warming, and there is no agreement either with a cooling trend from about 1940 in the north polar regions.

Indeed, the carbon dioxide effect on atmospheric temperature regimes is one of the major unsolved problems of modern climatology, though much basic information has accumulated since the IGY. Regular measurements have also been carried out in the Antarctic. With only a few interruptions, there is now a 25-year record for the South Pole 1957-81 (Keeling, 1960; Brown and Keeling, 1965; Keeling et al., 1976; Geophys. Mon., 1981 and 1982). In that time, the fraction of CO_2 in dry air has grown from 313 to 338 ppm. These values are only slightly smaller than, and the increase rate in close agreement with, the results of the Mauna Loa Observatory, Hawaii. In the Antarctic there is no vegetation which could originate seasonal variations of the concentration of CO_2. Notwithstanding, the mean annual range after removal of the linear trend is not zero but rather 1.6 ppm, about one-quarter of the value derived for Mauna Loa. All this is quite compatible with the notion of a net influx of air into the Antarctic late in spring and to a lesser amount late in fall. Such a flux of mass poleward must exist since the mean annual march of pressure, and thus mass, over the continent indicates the main maximum in summer, the secondary one in winter (see Section 6.4).

The worldwide increase of the atmospheric CO_2 concentration since the past century is very real, but the tentative conclusions regarding the possible consequences for the climate diverge (Flohn, 1980; Lindzen et al., 1982;

---

[*]A close relationship between temperatures at the coastal stations of East Antarctica and the duration of the presence of sea ice off shore as derived from satellite pictures was shown to exist by Budd (1975).

Elsaesser, 1983). Recent analyses of ice cores from the antarctic firn and ice sheet are of particular interest. Lorius (1983) found that prior to the present industrial age the CO_2 content of dry air may have been as low as 260 ppm, which is considerably lower than up to now generally assumed. More important still, Lorius discovered that about 10,000 years ago, "between Ice Age and Holocene", there appears to have been an increase of the CO_2 concentration comparable to the rise of the last hundred years. These new results could have a profound effect on present-day theories and expectations of global temperature variations in years to come.

Returning to the 78-year temperature record of ORCADAS, it cannot be denied that it contains more than only random effects, and the same is valid for the west side of the Antarctic Peninsula (Limbert, 1974) where the record of FARADAY has now grown to 36 years (Table A2). A comparison of the mean temperatures of early and recent decades and longer time intervals at ORCADAS and GRYTVIKEN is given in Table 6.4. A convincing explanation for a slight warming and the observed fluctuations has not yet been achieved. Many speculations can be found in the voluminous literature. Combinations of various processes were recently discussed by Damon and Kuenen (1976). As long, however, as the results of such possibly relevant ideas cannot be quantitatively ascertained and explained, there is not much satisfaction in considering the eventual consequences.

These critical comments on the interpretation of the march of temperature in one subantarctic region through three-quarters of the century have been made to give the reader a fair warning when now the much shorter annual temperature series of stations in higher southern latitudes will be looked at. The trends of four stations at different elevation and latitude are presented in Fig. 6.6, and decade values for a greater number of stations in Table 6.4. (Other series in Appendix A.2). It is interesting to note that at the SOUTH POLE the variation from year-to-year is considerably smaller than at the other high-latitude stations with a record of twenty years or more, and that the trend is zero, or insignificant by any measure. The latter cannot be said for FARADAY and HALLEY. Almost the same records have recently been analyzed by Mayes (1981) who finds a warming trend from 1958 to 1974 for a combination of various west antarctic stations, and from 1960 to 1980 for an east antarctic group. The statistical significance of the latter trend appears to be marginal. It was supported by the results of Thomas (1976) who made use of a fair number of ten-meter (below surface) temperature measurements in large parts of the Ross Ice Shelf during the IGY (1957-58) and again during the Ross Ice Shelf Research Project (1973-75). Thomas found an increase of temperature, as average for the Ross Ice Shelf between 150 and 170°W, of about 1° in 16 years, and

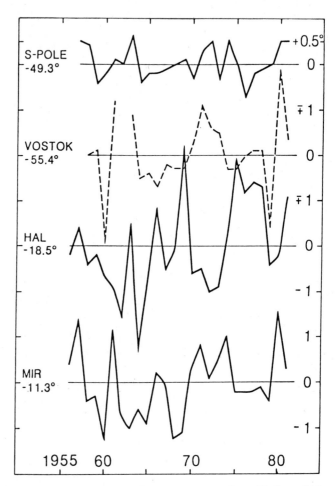

Fig. 6.6. Deviations of the annual temperatures from their multiannual averages, at S-POLE, VOSTOK, HALLEY, and MIRNY.

compared this with a similar increase of the annual mean air temperature of
BYRD Station and MCMURDO. However, BYRD's record of air temperatures ends with
February 1970, and Thomas' last measurements of the ice shelf temperature ten
meter below the surface was made in the summer 1974/75. Then, the unexpected
happened: at MCMURDO each of the next five years, 1975 to 1979, came out with
an annual temperature below the multiannual average, and together they made
the second coldest pentad on record. There can be no doubt that some minor ups
and downs can be found in many series of annual temperatures, but it remains
debatable whether such small modifications or oscillations really deserve the
somewhat pretentious and alarming name "climate variations".

TABLE 6.4. Comparison of early and recent decade means of annual temperatures.

Stat.	Decade	T̄	Decade	T̄	ΔT̄
SPO	1957-66	-49.3	1973-82	-49.3	+0.0
VOS	1958-68*	-55.6	1973-82	-55.4	+0.2
BEL	1955-64	-22.7	1970-79	-22.2	+0.5
MCM	1957-66	-17.7	1973-82	-17.5	+0.2
SCO	1958-67	-20.4	1973-82	-20.2	+0.2
HAL	1957-66	-19.0	1973-82	-17.9	+1.1
SAN	1957-66	-17.4	1973-82	-16.3	+1.1
NOV	1961-70	-10.9	1973-82	-10.3	+0.6
MAW	1955-64	-11.0	1973-82	-11.3	-0.3
MIR	1957-66	-11.6	1973-82	-11.2	+0.4
CAS	1957-66	- 9.7	1973-82	- 8.7	+1.0
FAR	1947-56	- 5.5	1973-82	- 4.4	+1.1
ESP	1952-61	- 5.9	1969-78	- 5.6	+0.3
ORC	1957-66	- 3.8	1973-82	- 4.0	-0.2
GRY	1957-66	+ 1.9	1972-81	+ 2.1	+0.2
MAC	1951-60	+ 4.6	1973-82	+ 5.0	+0.4
	first half (approx.)		second half of series		
ORC	1904-40	- 4.4	1941-82	- 4.0	+0.4
GRY	1905-40	+ 1.7	1941-81	+ 2.0	+0.3
	coldest decade		warmest decade		
ORC	1926-35	- 5.4	1955-64	- 3.3	2.1
GRY	1926-35	+ 1.4	1971-80	+ 2.1	0.7

*no data of VOS for 1962

6.3.2 Seasonal variations

While no significant variation of the annual temperature can be found in the 26 year record of the SOUTH POLE station*, there is some evidence for a growth of the difference between summer (NDJ) and winter (JJA) from the first to the second decade of measurements (van Loon and Williams, 1977). November has been chosen as the third summer month instead of February because the former shows a pronounced increase, the latter only an insignificant decrease.

---

*It must be hoped that the discrepancies between the monthly mean values obtained in recent years at the Clean Air Facility (published in Geophys. Mon., 1981 and 1982) and the values determined with the routine synoptic observations will soon be removed. The latter are the values regularly transmitted in the monthly CLIMAT reports, sometimes published in the "Monthly Climatic Data for the World" since 1957, and regularly in the Antarctic Journal of the U.S. since 1974.

The average temperatures of summer and winter amount to

-31.8 and -58.2° in the decade 1957-66,

-31.1 and -59.8   in the decade 1973-82.

    This suggests that in summer a clearer sky and hence stronger insolation, and in winter less cloudiness and therefore a more effective energy loss by out-going longwave radiation, could contribute to increase the mean range of winter to summer from 26.4 to 28.7°. Unfortunately, an adequate series of radiation measurements is not available, and the published cloud statistics of the sta-tion have been modified in 1970 so that a comparison of earlier and later years is not possible. Only the temperature series of VOSTOK lends support to con-sider the variation of the annual range on the Antarctic Plateau not as a random effect alone (Table 6.5). Obviously, such a variation cannot be a unidi-rectional development ad infinitum, and the data for the last pentad (1977-81) suggest just that. The other two stations in the Table have been added for a comparison only.

TABLE 6.5. Pentad mean temperature deviations from the overall seasonal average, in summer (November to January) and winter (June to August).

Pentad	Summer				Winter			
Stations	S POLE	VOS	HAL	MIR	S POLE	VOS	HAL	MIR
1957-61	-0.4	-0.5*	-0.5	-0.5	+0.5	-0.7*	-0.5	-0.2
1962-66	-0.6	-0.7*	-0.9	-0.5	+1.6	+0.8*	+0.7	-0.6
1967-71	+0.2	+0.3	+0.4	+0.1	-1.1	-0.7	-1.1	-0.4
1972-76	+0.7	+0.5	+0.5	+0.6	-0.7	+0.1	-0.4	+0.9
1977-81	+0.3	+0.6	+0.3	-0.0	+0.0	+0.6	+1.2	+0.8

*For VOSTOK only 1958-61 and 1963-66.

    A remarkable increase of annual, and most of all the fall and winter half-year, temperatures has occurred on the west side of the Antarctic Peninsula. This region is well represented by FARADAY Station, located on one of the "Argentine Islands", with its reliable, uninterrupted record since 1947. An examination of the variation of the seasonal mean temperatures from the 1950s to the 1970s makes it possible to speculate about the origin of the phe-nomenon. A comparison of the monthly mean values for the first thirteen years (1947-59) with the most recent ones (1970-82) shows that the temperature increase in the months April to September is definitely significant (ΔT greater than 2.0°), in the months November to February not (less than 0.6°). April to June is the season in which the sea southwest and west of FARADAY is covering with ice. When that happens early and extensively, low temperatures tend to prevail through the winter. A late ice cover could be the consequence of a diminished advection of cold air and ice from the southern part of the Bellingshausen Sea with its abundance of ice.

6.4 PERIODIC VARIATIONS OF PRESSURE

For an understanding of periodic variations of atmospheric pressure, it is helpful to make a mental picture of the physical process in progress when the pressure at a station changes with time. An increase means that the total mass of air in an (imaginary) vertical column above the barometer is augmenting, and vice versa. Such a change with time must be related to the three-dimensional motion of air over the station. If there were calm in all layers, there would be no change. This statement, however, is not reversible. Mass changes in different layers of the atmosphere can counterbalance. In fact, the tendency to compensation by differential advection together with divergent flow in one layer, convergent in another, is characteristic for the atmospheric circulation. What vertical extent of the atmosphere has to be taken into account depends mainly on the accuracy required. Since the atmospheric pressure at sea level is near 1000 mb (10^5 Pascal), at 20 km height about 55 mb, and at 50 km about 1 mb, the advection of different air masses in the mesosphere alone cannot contribute significantly to pressure changes observed at the Earth's surface.

These thoughts about the nature of pressure variations are pertinent because the yearly march of pressure on the Antarctic continent is characterized by a yearly and a half-yearly periodicity. These highly significant changes of pressure over a wide area permit conclusions regarding the net in- and outflow of air in different parts of the year. While the annual rhythm might have been expected in consideration of the pronounced summer vs. winter temperature contrast on the plateau, the strong semiannual period needs particular attention. When it was first made known (Schwerdtfeger, 1967), only nine years of data for SOUTH POLE and BYRD, and $5\frac{1}{2}$ years for VOSTOK, were available. At the time of this writing, it is 26 years for S-POLE, 23 for VOSTOK. Fig. 6.7 shows the mean monthly values of surface pressure at these two stations, as well as the first and the second harmonic and the sum of both only for VOSTOK. The percentage of the total variance accounted for by the sum of these two harmonics is 98% for S-POLE and 99% for VOSTOK.

Fig. 6.8 shows the areal extension of the most pronounced change, from October to December. The increase of mass inside the approximately circular 10 mb isoline corresponds to nearly 2% of the total mass over the continent. This means that between 1 and 2 × 10^{15} kg of air which in the southern spring

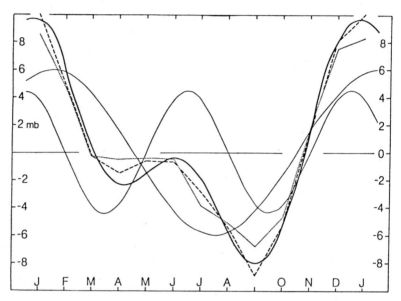

Fig. 6.7. Average annual march of pressure on the Antarctic Plateau. Dotted line: Deviation of the multiannual monthly values from the annual at the South Pole, 26 years, $\bar{p}$ = 681.3 mb.
Dashed line: Same for VOSTOK, 23 years, $\bar{p}$ = 624.8 mb.
The two thin solid lines indicate the annual and the semiannual components of the VOSTOK series. The medium solid line shows the sum of these two harmonics that correspond to 62% and 37% of the total variance, respectively.

must have been at lower latitudes, are in December and January much closer to the Earth's axis of rotation. It implies that the angular momentum of the planet must be affected, so that the length of day has to <u>decrease</u> slightly -- but measurably -- in this part of the year and still more slightly <u>increase</u> in the next months, as Fig. 6.7 suggests (Munk and MacDonald, 1960; Frostman et al., 1968). The concept that a net transport of 10^{15} kg of air across a circular line like the 70°S parallel can really occur within the time interval of two months, is entirely realistic. If the flux were continuous and constant through troposphere and lower stratosphere, the N to S component of the winds would have to be not more than 10^{-3} m/sec.

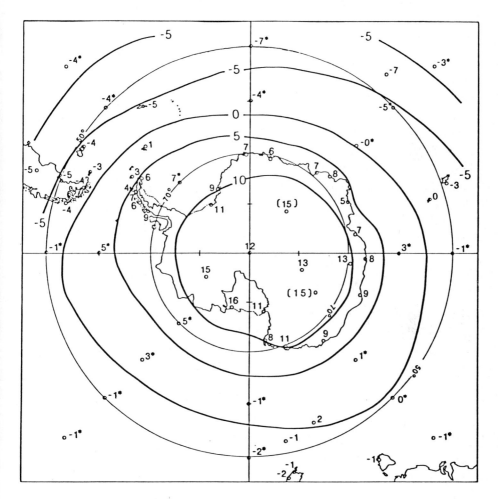

Fig. 6.8. Multiannual average change of surface pressure (in millibar) from October mean to December mean. The two numbers in parentheses are based on three year records only.

6.5 A BRIEF COMPARISON OF SOUTH- AND NORTH-POLAR CLIMATES

The difference, - one might even say the contrast - between the climates of South and North is essentially due to the diversity of the surfaces and the topographic characteristics. An anonymous author (1978) has expressed that in the most concise form: "The south-polar region is an ice-covered continent surrounded by ocean; the north-polar region an ice-covered ocean surrounded by continents." The consequences are manifold; they affect i: the atmospheric circulation.

a) There is no all-year-round sea level low pressure trough and corresponding windfield near the polar circle of the Northern Hemisphere, as it is in the Southern.

b) The equinoctial maxima of the westerly winds in the lower troposphere of middle and subpolar latitudes are pronounced in the south, negligible in the north.

c) The average circumpolar westerly winds in the upper troposphere are in the southern summer as strong as in the northern winter (van Loon, 1972; Flohn, 1978).

d) Drastic changes of the atmospheric circulation, leading to strong sudden stratospheric warmings and reaching down into the lower part of the stratosphere, have occurred in the northern winter a few times in every decade since 1951, but have not appeared, at least since 1957, in the southern winter.

ii: Moisture content and temperature regime.

a) Since the moisture content of the air (unit: specific humidity q in kg H_2O vapor/kg air) at low temperatures can only be small (Table 5.1), the air over the high plateau of Antarctica cannot contain much H_2O. In comparison, the moisture transport toward the high northern latitudes is much less restricted by the topography; in fact, it is favored in the Atlantic sector where the poleward flow of relatively warm water through the Greenland – Spitzbergen Passage supplies energy continuously. It is obvious that the poleward decrease of the moisture content in the lower troposphere must be much greater in the South.

b) This shortage of H_2O in the air over the plateau directly diminishes the cloudiness, correspondingly affecting the radiation budget and the temperature regime. The latter becomes evident in Fig. 6.9, contrasting the mean summer and winter temperatures in troposphere and lower stratosphere near the parallels 80°S and 80°N. The graphs show that in the 600 to 400 mb layer, i.e. nearly everywhere above the boundary layer, the difference "northern minus southern temperatures" is much larger for the summer- than for the winter-atmospheres. That fits well with the fact that the winter to summer increase of cloudiness is much greater in the Arctic than over the Antarctic Plateau (Vowinckel and Orvig, 1970; Schwerdtfeger, 1970).

c) Regarding the short polar summer alone, it is of interest that on the Antarctic Plateau, and also at two stations near sea level at 78°S, December and January have about the same mean monthly temperature (see Table A3). In contrast, the nine weather stations on firm ground and eight "North Pole" floating ice station series listed by Vowinckel and Orvig (1970), all north

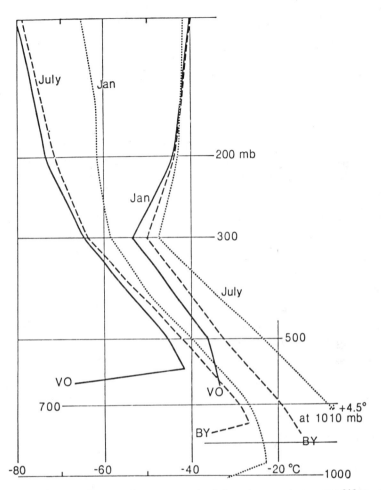

Fig. 6.9. Comparison of mean January and July temperature profiles near 80°S and N.
Solid lines: VOSTOK, 78.5°S, 3488 m; dashed lines: BYRD, 80°S, 1530 m.
Dotted lines: Average of three northern stations,
EUREKA, 80.0°N, 85.9°W, 10 m
NORD, 81.6   16.7   35
ALERT, 82.5   62.3   67   .

of 78°N, have from June to July a temperature increase averaging more than 2°.

d) Under the hard conditions of the Northeast-Siberian winter, the surface inversion can become as large as it is over the Antarctic Plateau, as Table 6.6 shows. However, the duration of such an extremely stable boundary layer is quite different: On the plateau, the inversion remains that strong

TABLE 6.6. Comparison of the average surface temperature inversions in the winter on the East Antarctic Plateau and in Northeastern Siberia.

VOSTOK, 3488 m   VERKHOYANSK, 137 m

78.5°S, 106.8°E   67.6°N, 133.4°E
July 1964 - 1970   January 1974 - 1981
624 mb, surface: -67.0°   1009 mb, surface: -44.5°
500 "   4930 m: -45.4°   850 "   1350 m : -23.2°
diff.: 21.6° / 1440 m   diff.: 21.3° / 1210 m

for more than 200 days per year, while in a North-Siberian valley it is less than 100 days. In the latter, there is no such phenomenon like a coreless winter (Fig. 6.10).

e) Finally, a comparison of the strength of the winter in subpolar latitudes, like southern ocean versus the northern coast of European Russia, might be of interest. For large scale climatological considerations it would be, of course, much more valuable to study the temperature regimes of wide areas, but one would have to wait some fifty more years to get appropriate series of data for the southern latitudes. Fletcher (1969) compared the winter (Jan. + Feb.) at ARCHANGEL with the month of August at ORCADAS; he chose the latter because there is generally more ice around ORCADAS than earlier in the year. Fletcher's comparison ended with the decade 1952-1961. It is now possible to add 21 years. It appears preferable to look for both places not at one individual month, but rather the three coldest ones usually called winter. It is remarkable how much the overall picture has changed in the last twenty years. Fig. 6.11 can easily serve as a fair warning against premature conclusions regarding minor climatic changes and their interrelations.

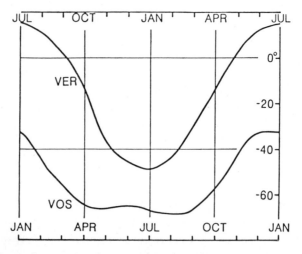

Fig. 6.10. Comparison of the average annual march of temperatures at VERKHOYANSK, NE-Siberia, and VOSTOK, High Plateau.

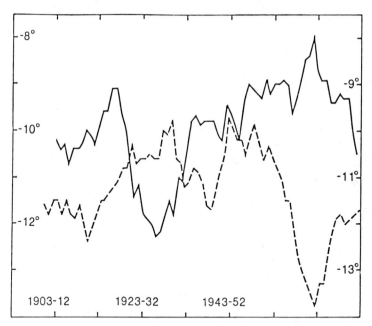

Fig. 6.11.   Overlapping   ten-year   averages   of air temperature in the winter:
ORCADAS,   June   to   August   from   1903-12 to 1972-81, solid line.
ARCHANGEL, December to February from 1901-10 to 1973-82, dashed line.

iii: The seasonal variation of the ice cover.

In the majority of all years, the ice cover of the arctic waters reaches
the maximum size in March, a bit less than twice the minimum of September. In
the south-polar regions, the variation in size is much greater. In September
the sea ice belt around the continent is about six to eight times as large as
the covered area in March. There are several reasons for such dissimilar variations. One of the causes certainly is that the Southern Hemisphere sea ice lies
in, or moves toward, comparably low latitudes. Another reason must be that the
Sun-Earth distance is smallest in the southern summer. A third one is the frequent occurrence of intense storms over the southern ocean in the southern
spring; they help to destroy ice fields which could survive under less exacting
weather conditions. Again, the concepts of weather and climate are inseparable.

APPENDIX

TABLE A 1 Meteorological stations in the Antarctic, either with records of two winters at least, or in operation in 1983. Symbols: * = active in 1983; i = with interruptions; AWS = automatic weather station.

Name	Abbr.	Lat.	Long.	El.(m)	in operation

a) Interior

AMUNDSEN-SCOTT*	SPO	90°		2835	since Jan. 1957
BYRD	BYR	80.0	120.0 W	1530	Jan. 1957 - Dec. 1970; AWS since 1980
PLATEAU ST.	PLA	79.2	40.5 E	3625	Dec. 1965 - Jan. 1969
VOSTOK*	VOS	78.5	106.9 E	3488	1958 - 1961 and since 1963
VANDA	VAN	77.5	161.6 E	94	1969, 70, 74; later years summer only
SIPLE	SIP	75.9	84.2 W	1050	1978 - 1982
EIGHTS	EIG	75.2	77.2 W	420	Dec. 1962 - Oct. 1965
DOME C (AWS)*	DOM	74.5	123.0 E	3280	Feb. 1980 - Jan. 1983
MIZUHO*	MIZ	70.7	44.3 E	2230	since Apr. 1976, earlier i
PIONERSKAYA	PIO	69.7	95.5 E	2740	May 1956 - Dec. 1958

b) Around the continent (Antarctic Peninsula follows under c))

ELLSWORTH	ELL	77.7	41.1 W	42	Mar. 1957 - Dec. 1962
BELGRANO 1)	BEL	78.0	38.8 W	50	Feb. 1955 - Sep. 1979
SHACKLETON	SHA	78.0	37.2 W	58	Mar. 1956 - Oct. 1957
HALLEY*	HAL	75.5	26.6 W	35	since Feb. 1956
MAUDHEIM	MAU	71.0	10.9 W	38	Mar. 1950 - Jan. 1952
NEUMAYER*	NEU	70.6	8.2 W	40	since Mar. 1981
NORWAY ST.	NOR	70.5	2.5 W	55	1957 - 1961
SANAE*	SAN	70.3	2.4 W	52	since 1962 (62 i) } series combined
NOVOLAZAREZ*	NOV	70.8	11.8 E	99	since Feb. 1961
BAUDOUIN	BAU	70.4	24.3 E	37	1958 - 1960, 1964 - 1966 i
SYOWA*	SYO	69.0	39.6 E	15	1957, 1959 - 1961, since Feb. 1966
MOLODEZNAYA*	MOL	67.7	45.9 E	39	since Feb. 1963
MAWSON*	MAW	67.6	62.9 E	8	since Feb. 1954
DAVIS*	DAV	68.6	78.0 E	12	1957 - 1964, since 1969
MIRNY*	MIR	66.6	93.0 E	42	since Mar. 1957
OASIS	OAS	66.3	100.7 E	28	Nov. 1956 - Oct. 1958
CASEY*	CAS	66.3	110.5 E	12	since Mar. 1957, first years WILKES
DUMONT D'URV.*	DUD	66.7	140.0 E	40	since Jan. 1956
PORT MARTIN	PMA	66.8	141.4 E	14	Feb. 1950 - Jan. 1952
CAPE DENISON	DEN	67.1	142.7 E	6	Feb. 1912 - Dec. 1913
LENINGRADSKA*	LEN	69.5	159.4 E	295	since Mar. 1971
CAPE EVANS	CEV	77.6	166.4 E	20	Jan. 1911 - Dec. 1912
HUT POINT	HUP	77.9	166.7 E	12	Feb. 1902 - Jan. 1904
MCMURDO*	MCM	77.9	166.6 E	24	since Apr. 1956
SCOTT BASE*	SCO	77.9	166.7 E	14	since Mar. 1957
CAPE ADARE	CAD	71.3	170.2 E	6	Feb. 1899 - Jan. 1900; 1911
HALLETT	HTT	72.3	170.3 E	5	Feb. 1957 - Feb. 1964, + some summers
LITTLE AMERICA	LAM	78.6	163.9 W	40	1929/30, 34/35, 40/41, 1956 - 1958
FRAMHEIM	FRM	78.6	163.6 W	40	Feb. 1911 i - Jan. 1912
RUSSKAYA*	RUS	74.8	136.9 W	100	since 1980

----

1) Since 1980 there is a station BELGRANO II at 77°52' S, 34°34' W, 256 m.

Name

c) Antarctic Peninsula and islands (selection)

FOSSIL BLUFF	FOB	71.3	68.3 W	68	1968, several other years summer only
SAN MARTIN 2)*	SMA	68.1	67.1	4	since 1946 with many i
ADELAIDE ISL.	ADE	67.8	68.9	26	1962 - 1975
HORSESHOE ISL.	HOR	67.8	67.3	9	1955 - 1960
ROTHERA*	ROT	67.6	68.1	15	since 1976
FARADAY*	FAR	65.3	64.3	11	since Jan. 1947 (formerly ARG.ISL.)
MATIENZO	MAT	65.0	60.1	32	1962 - 1972, 1975, + several summers
ALT. BROWN*	ABR	64.9	62.9	7	1951 - 1959, since 1965 i
PALMER ST.*	PAL	64.8	64.1	8	since 1965
SNOW HILL*	SNO	64.4	57.0	13	Mar. 1902 - Oct. 1903
MELCHIOR	MEL	64.3	63.0	8	1947 - 1960
MARAMBIO*	MAR	64.2	56.7	198	since Oct. 1970
PRIMAVERA*	PRI	64.2	61.0	50	since 1979
PETREL	PET	63.5	56.3	18	Feb. 1967 - Dec. 1976; 1977 i
ESPERANZA 3)*	ESP	63.4	57.0	13	since 1945 i
O'HIGGINS*	OHI	63.3	57.9	10	since IGY, few data published
DECEPTION 4)	DEC	63.0	60.7	8	1944 - 1967, except Jan. 46 - Jan. 47
PRATT*	PRA	62.5	59.7	5	since IGY, few data published
MARSH*	RMA	62.2	58.9	10	since 1960 (formerly P.FREI)
BELLINGSHAUSEN*	BHN	62.2	59.0	14	since 1968
ADMIR.BAY	ADB	62.1	58.4	9	1948 - 1960
ARCTOWSKI*	ACT	62.1	58.5	x	since 1980
SIGNY ISL.*	SIG	60.7	45.6	7	since 1947
ORCADAS*	ORC	60.7	44.7	4	since Apr. 1903

d) Islands between 54 and 60°S

COR URUGUAY	CUR	59.5	27.3 W	14	1979 and 1980
USHUAIA*	USU	54.8	68.3 W	7	since 1931
MACQUARIE ISL.*	MAC	54.5	158.9 E	30	since 1949
GRYTVIKEN*	GRY	54.3	36.5 W	3	since 1905, 1982 i

e) Other stations mentioned in the text

SOUTH ICE		82.0	28.9 W	1350	1957 i
SOVIETSKAYA		78.4	87.5 E	3660	Feb. - Dec. 1958
KOMSOMOLSKAYA		74.1	97.5 E	3500	Nov. 1957 - Feb. 1959
CHARCOT		69.4	139.0 E	2400	Apr. 1957 - Dec. 1958
FRAMHEIM		78.6	163.6 W	40	Feb. 1911 i - Jan. 1912

- - - - - - - - - - - - -

2) at or near Stonington Island.
3) some years data from neighboring station HOPE BAY, some years from both.
4) Data from Argentine and British stations.

230

TABLE A2 Annual mean temperature at stations with 20 or more years of record.
On this page all values are negative °C. The asterisks indicate that data for
one month are missing and have been substituted by the multiannual monthly
average.

	SPO	VOS	BEL	HAL	SAN	NOV	SYO	MOL	MAW	MIR	CAS	DUD	MCM	SCO
1955	-	-	20.4	-	-	-	-	-	10.6	-	-	-	-	-
56	-	-	22.1	18.7*	-	-	-	-	10.8	10.9*	-	10.5	-	-
	48.8	-	21.9	18.1	17.0*	-	10.1*	-	10.0	9.9	8.1	10.8	17.3	
58	48.9	55.4	23.5	18.9	16.6	-	-	-	11.2	11.8	9.7	11.4	18.5	21.2
	49.7	55.3	21.7	18.7	16.2	-	10.8	-	10.8	11.6	9.8	11.8	18.2	20.9
60	49.5	57.2	22.7	19.2	18.7	-	11.8	-	12.5	12.6	10.5	11.2	18.2	21.6
	49.2	54.2	22.7	19.4	18.1	10.3*	10.5	-	9.7	10.3	8.3	10.9	17.0	19.8
62	49.3	-	24.1	20.1	17.8*	-	-	-	12.6	12.3	9.8	10.3	18.5	21.8
	48.7	54.5*	21.8	18.0	17.3	11.2	-	10.9*	10.7	12.4	10.7	11.6	17.7	20.4
64	49.7	55.9	25.9	20.8	17.8	10.8	-	10.7	10.9	12.0	10.7	10.9	17.5	19.7
	49.5	55.7	21.9	19.2	17.9	11.1	-	11.4	12.1	12.3	10.0	10.8	17.4	20.2
66	49.5	56.1	21.8	17.7	16.5	11.0	11.0*	11.5	11.1	11.1	9.5	10.7	16.4	19.5
	49.4	55.6	22.6	19.0	16.7	9.9	10.2	10.1	10.8	11.0	8.7	10.7	16.9	19.3
68	49.3	55.7	21.0	18.6	18.2	11.0	10.2	10.7	11.4	11.4	9.1	9.8	18.3	21.1
	49.2	55.7	20.1	16.3	16.6	10.7	9.6	10.8	11.5	12.4	11.1	11.9	17.7	20.7
70	49.6	54.9	20.9	19.1	18.2	10.8	10.6	10.9	11.1	11.0	10.1	11.1	15.4	19.0
	49.0	54.5	23.2	19.0	18.7	10.8	10.9	11.2	10.7	10.4	8.4	10.0	17.2	19.1
72	48.8	54.8	22.8	19.5	17.8	10.3	10.5	10.7	10.6	11.2	9.0	10.0	15.5	18.5
	49.6	54.9	22.9	19.4	16.7	10.4	10.7	11.3*	11.0	10.5	8.5	10.4	16.9	19.2
74	48.8	55.7	22.9	18.4	14.6	10.1	10.5	10.4	10.3	10.3	8.8	10.7	17.0	18.9
	49.3	55.7	21.8	16.6	14.0	10.4	11.4	11.4	11.0	11.5	9.2	11.0	17.6	19.2
76	50.0	55.4	21.0	17.3	16.9	11.4	12.1	12.0	11.8	11.5	9.3	11.3	18.4	22.3
	49.5	55.3	22.3	17.1	15.8	10.0	10.0	10.4	10.5	11.5	8.7	10.3	17.7	20.9
78	49.4	55.3	21.3	17.2	17.4	10.3	10.7	10.6	11.1	11.4	9.2	10.4	18.5	21.3
	49.3	57.0	-	18.9	18.1*	10.5	10.9	11.2	11.9	11.7	8.9	10.7	17.9	20.7
80	48.8	53.5	-	18.7	18.0*	9.3	8.2	10.9	10.8	9.8	6.5	9.3	16.3	18.6
	48.8	55.1	-	17.4	15.1	9.9	9.7	11.0	11.4	11.0	7.3	8.8	16.8	19.6
82	49.2	55.9	-	17.6	16.7	10.7	-	11.5	13.4	12.7	10.3	10.3	18.0	21.0
Av.	49.3	55.4	22.2	18.5	17.1	10.6	10.5	11.0	11.2	11.4	9.2	10.7	17.4	20.2

TABLE A2, continued

year	ORC	GRY	USU		year	ORC	GRY	USU	MAC	FAR	ESP
1904	-5.4	–			1944	-4.0	1.9	5.8			
	-4.6	2.1				-7.1	1.2	6.1			
06	-4.6	2.3			46	-3.6	1.4	5.3			-5.3
	-4.9	2.1				-3.9	2.3	5.5		-4.1	-5.8
08	-3.0	2.3			48	-4.9	1.4	4.9		-5.9	-7.1
	-4.5	2.0				-6.0	2.2	5.5		-6.9	–
10	-3.4	2.1			50	-4.8	0.8	4.9		-5.3	–
	-3.8	1.4				-3.3	1.8	5.8	4.5	-4.4	–
12	-5.8	1.4			52	-4.1	1.7	6.0	4.6	-4.9	-5.6
	-5.0	1.2				-3.9	2.0	6.0	4.6	-7.5	-5.8
14	-4.0	2.2			54	-4.2	1.9	5.8	4.1	-5.9	-6.7
	-6.3	1.6				-2.2	2.4	6.0	4.6	-3.6	-4.7
16	-5.5	1.6			56	-1.4	2.5	6.5	5.1	-1.9	-4.2
	-3.3	2.4				-3.5	2.4	5.5	4.8	-4.8	-5.5
18	-3.8	2.3			58	-5.0	1.9	5.1	4.6	-7.5	-6.7
	-3.5	1.9				-5.4	1.2	4.7	4.2	-8.1	-7.1
20	-4.7	1.2			60	-2.9	2.1	5.6	4.8	-5.1	-5.5
	-4.6	1.4				-4.3	1.8	4.9	4.3	-4.6	-7.4
22	-4.2	2.3			62	-2.1	2.5	6.2	4.9	-3.7	-3.5
	-4.2	1.4				-3.6	1.9	5.3	4.0	-5.2	-5.3
24	-5.1	1.3			64	-4.0	1.3	5.4	4.5	-4.6	-6.9
	-4.5	1.7				-2.9	2.3	5.7	4.7	-4.3	-5.5
26	-4.4	1.0			66	-4.7	1.6	5.4	4.4	-5.2	-5.6
	-5.5	0.8				-4.0	1.9	5.6	4.7	-4.4	-6.3
28	-6.7	1.2			68	-3.1	2.5	6.0	4.9	-3.5	-5.6
	-5.5	1.7				-3.9	2.3	5.4	4.4	-5.0	-6.3
30	-7.0	1.3			70	-3.4	2.1	5.8	4.7	-3.2	-4.6
	-4.5	1.8	5.7			-3.5	2.1	5.2	5.6	-1.9	-5.4
32	-5.1	1.7	5.0		72	-5.1	1.6	5.4	4.4	-2.3	-6.8
	-5.1	1.7	5.6			-4.7	2.2	5.2	4.7	-3.5	-6.9
34	-4.6	1.7	5.8		74	-4.6	2.2	5.9	4.9	-2.1	-5.9
	-5.4	0.7	5.0			-4.8	2.3	5.5	4.9	-3.4	-7.3
36	-3.6	2.1	5.8		76	-3.2	2.0	5.2	4.9	-5.0	-5.1
	-2.9	1.9	5.1			-2.8	2.0	5.7	5.1	-5.1	-5.4
38	-4.9	1.7	5.0		78	-3.2	2.4	6.0	5.2	-5.4	-4.4
	-5.8	0.6	4.7			-3.8	2.4	6.2	5.1	-4.3	–
40	-3.8	2.3	5.5		80	-6.1	1.7	5.4	5.5	-6.1	-7.5*
	-4.2	1.8	5.6			-3.9	2.1	5.8	5.2	-4.8	–
42	-5.1	2.2	5.8		82	-3.1	–	5.9	4.7	-3.9	–
43	-2.6	2.8	6.1								
					Av.	-4.3	1.8	5.5	4.7	-4.7	-5.8

TABLE A3, T̄: Monthly and annual average temperature at stations with ten or more years of record, and a few selected places of shorter activity. The number of years of data used follows the abbreviated name of the station; coordinates are given in Table A1. For the temperature values, no sign means below zero °C.

STAT.	Ys	J	F	M	A	M	J	J	A	S	O	N	D	Year

a) Interior

STAT.	Ys	J	F	M	A	M	J	J	A	S	O	N	D	Year
SPO	26	27.9	40.2	54.3	57.3	57.3	58.2	59.9	59.7	59.4	50.7	38.4	27.7	49.3
BYR	14	14.7	19.8	27.7	29.7	33.0	34.1	35.6	36.7	36.6	30.2	21.4	14.4	27.9
PLA	3	33.9	44.4	57.2	65.8	66.4	69.0	68.0	71.4	65.0	59.5	44.4	32.3	56.4
VOS	24	32.3	44.3	58.0	64.9	65.9	65.1	67.0	68.3	66.3	57.1	43.4	32.3	55.4
VAN	3	+1.3	5.9	20.4	29.7	29.2	30.0	38.0	32.3	31.2	15.7	6.2	+0.2	19.8
SIP	5	11.9	18.7	24.9	29.0	27.1	30.9	30.5	36.7	30.3	26.6	17.7	13.1	24.8
EIG	3	10.0	18.0	25.0	31.1	33.1	33.5	33.5	37.0	34.3	28.3	17.3	11.3	26.0
DOM	3	30.0	41.0	54.3	61.1	61.4	58.5	58.5	61.7	57.8	51.9	39.3	30.0	50.5
MIZ	6	18.6	24.3	31.5	36.5	38.7	40.9	39.1	40.4	39.3	35.2	25.0	18.0	32.3
PIO	2	23.4	32.4	38.4	39.2	42.9	46.0	47.3	48.3	44.5	39.3	31.4	23.1	38.0

b) Around the continent (Antarctic Peninsula follows under c)

STAT.	Ys	J	F	M	A	M	J	J	A	S	O	N	D	Year
ELL	6	8.3	15.3	22.9	27.4	28.2	32.2	33.3	33.3	30.3	21.9	14.4	7.8	22.9
BEL	24	6.3	13.2	22.0	26.2	29.6	32.0	33.5	32.9	31.0	21.7	12.7	6.1	22.2
HAL	27	4.8	9.8	16.6	20.0	24.4	26.6	28.9	28.5	26.2	19.2	11.6	5.3	18.5
SAN	25	4.1	8.9	14.5	19.3	21.2	23.4	27.1	27.2	25.2	18.8	10.8	5.2	17.1
NOV	22	0.7	3.7	8.0	12.1	13.3	15.9	17.7	18.5	17.4	12.9	5.9	1.1	10.6
BAU	5	4.5	8.3	12.0	16.1	17.6	20.3	23.1	23.5	23.5	17.6	10.7	5.6	15.2
SYO	20	0.6	3.3	6.4	9.9	13.4	15.8	18.0	19.3	18.3	13.3	6.4	1.5	10.5
MOL	20	0.5	4.0	8.3	11.4	14.2	16.3	18.0	18.7	17.8	13.8	6.6	1.4	10.9
MAW	28	+0.1	4.4	10.3	14.5	16.1	16.8	17.8	18.8	17.7	13.2	5.4	0.3	11.3
DAV	19	+0.6	2.6	9.3	13.1	15.7	15.6	17.3	17.1	16.5	12.5	5.1	0.0	10.3
MIR	27	1.6	5.3	10.1	13.9	15.3	15.6	16.6	17.4	17.2	13.5	7.0	2.4	11.3
CAS	26	+0.1	2.3	6.4	11.4	13.4	14.3	14.6	15.1	14.8	11.2	5.4	0.9	9.2
DUD	27	0.7	4.2	8.7	12.7	14.7	16.0	16.2	16.8	16.1	13.2	7.0	1.7	10.7
PMA	2	1.6	5.5	10.7	15.8	16.0	19.3	19.0	18.3	19.1	13.0	5.9	2.5	12.2
DEN	2	0.9	4.4	10.9	16.6	18.2	20.2	19.6	18.0	18.4	15.0	8.2	2.9	12.8
LEN	11	4.0	7.4	10.8	15.2	18.4	19.4	20.8	21.3	21.0	18.9	10.8	4.9	14.4
MCM	26	3.1	8.8	17.6	21.1	23.3	23.5	25.8	26.9	25.0	19.5	9.9	3.8	17.4
SCO	25	4.9	10.7	20.3	24.4	27.2	26.3	29.3	30.5	28.4	22.7	11.9	5.3	20.2
HTT	8	1.1	3.2	10.5	17.8	22.6	23.0	26.4	26.6	24.5	18.3	8.0	1.7	15.3
LAM	6	6.6	12.9	21.8	28.3	30.6	27.9	36.3	36.6	36.7	25.4	15.6	6.4	23.8

TABLE A 3 T, continued

STAT.	Ys	J	F	M	A	M	J	J	A	S	O	N	D	Year
							c) Antarctic Peninsula							
SMA	15	+0.4	1.3	3.1	6.2	9.4	13.1	14.4	13.2	10.4	7.3	3.3	+0.1	6.8
ADE	14	+0.6	0.0	1.5	4.6	6.5	8.8	9.7	11.4	8.1	5.3	2.2	+0.2	4.8
ROT	6	+0.7	0.4	2.1	4.7	7.8	9.8	15.3	14.0	11.0	8.3	4.2	0.1	6.4
FAR	36	+0.3	+0.1	1.0	3.4	5.8	8.1	10.7	11.0	8.3	5.1	2.6	0.4	4.7
MAT	12	1.5	5.0	11.3	16.4	19.2	21.7	20.5	20.6	15.0	9.7	5.7	2.1	12.4
ABR	21	+1.6	+1.2	0.2	2.2	4.4	6.2	8.3	7.2	5.4	3.0	0.5	+0.8	2.8
PAL	6	+1.4	+1.1	0.2	2.4	4.0	6.4	10.8	11.4	7.6	5.3	1.7	+0.5	3.9
SNO	2	0.9	3.6	10.3	13.8	18.2	19.7	20.8	19.2	15.6	9.5	8.1	2.0	11.8
MEL	14	+1.1	+0.5	0.7	2.6	4.8	7.3	9.2	8.7	6.2	3.2	1.5	+0.2	3.7
MAR	11	1.9	3.3	6.8	12.3	12.9	14.9	16.3	16.2	12.7	6.4	4.3	1.9	9.2
PET	10	0.5	1.3	4.9	9.7	12.2	13.3	12.1	12.5	10.2	3.5	1.8	0.2	6.9
ESP	31	+0.2	1.3	3.7	7.5	9.3	11.4	11.8	10.6	7.4	4.1	2.1	0.2	5.8
OHI	12	+0.5	+0.1	1.2	3.8	6.0	7.9	9.2	9.1	6.6	3.2	2.2	+0.1	4.0
DEC	23	+1.4	+1.1	+0.1	2.1	4.3	6.3	8.0	7.7	4.8	2.4	1.0	+0.5	2.8
PRA	10	+0.8	+1.4	0.1	1.5	3.2	5.4	7.7	8.2	5.3	2.2	1.3	+0.4	2.7
RMA	12	+1.3	+1.5	+0.3	1.1	3.4	5.4	7.6	8.0	5.1	2.5	1.0	+0.6	2.5
BHN	15	+1.1	+1.2	+0.1	1.8	4.0	6.0	7.1	7.3	4.6	2.6	1.2	+0.4	2.6
ADB	13	+1.3	+1.3	+0.1	2.4	4.9	6.5	8.5	7.4	4.4	1.4	0.8	+0.7	2.7
SIG	33	+0.8	+1.0	+0.1	2.2	6.5	8.8	10.0	9.3	5.4	2.3	1.3	0.0	3.7
ORC	78	+0.3	+0.5	0.6	3.0	6.7	9.8	10.5	9.8	6.4	3.4	2.1	0.5	4.3
						d) Islands between 54 and 60°S								
CUR	2	0.0	+0.2	0.5	2.2	7.3	11.2	12.1	9.6	7.0	4.9	2.5	0.9	4.8
USU	52	+9.1	+9.0	+7.8	+3.2	+1.7	+1.7	+1.4	+2.0	+3.8	+5.9	+7.2	+8.5	+5.5
MAC	33	+6.9	+6.8	+6.3	+5.2	+4.2	+3.3	+3.2	+3.3	+3.4	+3.8	+4.5	+6.1	+4.7
GRY	77	+4.7	+5.4	+4.6	+2.6	+0.2	1.2	1.5	1.5	+0.1	+1.8	+3.0	+3.8	+1.8

TABLE A3, Radiation: Mean monthly and yearly sums of radiative energy fluxes, in MJ m^{-2}. G = global radiation, a = albedo in percent, E_S = effective short-wave radiation, E_L = effective longwave radiation, NR = net radiation, σ_{NR} = standard deviation of NR.
Sources: Farman and Hamilton, 1978; Dolgina et al., 1976; JARE Reports 1981 and 1982; Thompson et al., 1971.

	J	F	M	A	M	J	J	A	S	O	N	D	Year
				HALLEY, 75.5°S, 30 m, 1963 - 72								
G	762	395	189	28	0.1	0	0	10	115	384	687	885	3455
a	81	83	83	82	-	-	-	80	81	82	83	81	% -
E_S	148	66	33	5	<0.1	-	-	2	22	68	120	166	630
E_L	-106	-65	-72	-54	- 51	-57	-60	-59	-70	-93	-100	-115	- 903
NR	42	1	-39	-49	- 51	-57	-60	-57	-48	-25	20	51	- 273
σ_{NR}	29	11	17	6	8	9	11	11	13	16	20	25	80

TABLE A3, Radiation, continued

	J	F	M	A	M	J	J	A	S	O	N	D	Year
FARADAY, 65.3°S, 11 m, 1963-72													
G	552	349	185	73	20	4	12	67	187	407	556	654	3065
a	69	65	70	74	75	79	77	80	82	84	81	74 %	-
E_s	171	123	56	19	5	1	3	14	34	67	106	169	767
E_L	-89	-86	-78	-68	-79	-65	-76	-72	-67	-76	-79	-106	-941
$\overline{NR}$	82	37	-22	-49	-74	-64	-73	-58	-33	- 9	27	63	-174
σ_{NR}	34	36	23	11	14	7	10	12	11	10	13	20	97
MIRNY, 66.6°S, 40 m, 1956-58, 63-73													
G	843	519	302	107	23	3	10	64	227	472	778	968	4316
a	78	81	82	85	87	-	-	86	86	85	84	81 %	-
E_s	185	97	53	16	3	1	2	9	32	70	123	183	774
E_L	-103	-67	-79	-74	-89	-75	-86	-79	-73	-75	-79	-119	-998
$\overline{NR}$	82	30	-26	-58	-86	-74	-84	-70	-41	- 5	44	64	-224
σ_{NR}	27	33	20	17	16	17	19	19	14	20	27	50	98
MOLODEZNAYA, 67.7°S, 37 m, 1963-73, i													
G	829	496	257	75	11	0	5	57	217	502	775	1004	4228
a	19	23	40	49	50	-	-	61	64	57	39	20 %	-
E_s	673	383	155	38	6	0	2	22	79	218	474	807	2857
E_L	-246	-151	-84	-61	-75	-77	-86	-84	-81	-87	-137	-273	-1442
$\overline{NR}$	427	232	71	-23	-69	-77	-84	-62	- 2	131	337	534	1415
NOVOLAZAREZKAYA, 70.8°S, 99 m, 1963-1973, i													
G	832	485	248	69	4	-	-	34	175	454	721	915	3937
a	20	21	25	26	-	-	-	33	34	29	24	22 %	-
E_s	666	386	187	51	3	-	-	23	116	325	548	717	3022
E_L	-264	-176	-143	-127	-121	-109	-123	-134	-137	-169	-204	-237	-1944
$\overline{NR}$	402	210	44	-76	-118	-109	-123	-111	- 21	156	344	480	1078
VANDA, 77.5°S 94 m, 1969 and 1970, i													
G	812	408	88	9	-	-	-	4	39	298	701	800	3159
a	20	20	43	43	-	-	-	43	28	27	22	20 %	-
E_s	650	326	50	5	-	-	-	2	28	218	547	640	2466
E_L	-353	-233	-42	-83	-131	-117	-133	-112	-116	-218	-307	-336	-2181
$\overline{NR}$	297	93	8	-78	-131	-117	-133	-110	- 88	0	240	304	285
MIZUHO, 70.7°S 2230 m, 1979 and 1980													
G	902	561	319	78	5	-	<0.5	38	205	521	847	1123	4599
a	81	81	81	76	-	-	-	79	80	81	79	79 %	-
E_s	172	105	61	19	1	-	<0.1	8	40	100	181	238	925
E_L	-154	-113	-95	-68	-70	-67	-88	-87	-110	-141	-174	-201	-1368
$\overline{NR}$	18	-8	-34	-49	-69	-67	-88	-79	- 70	- 41	7	37	- 443
VOSTOK, 78.5°S 3488 m 1963 - 1973, i													
G	1087	601	224	18	-	-	-	3	98	458	940	1235	4664
a	83	84	85	86	-	-	-	-	89	86	85	83 %	-
E_s	185	96	34	3	-	-	-	-	11	64	141	210	744
E_L	-131	-71	-63	-45	-46	-46	-46	-44	-49	-68	-102	-120	-831
$\overline{NR}$	54	25	-29	-42	-46	-46	-46	-44	-38	- 4	39	90	- 87

TABLE A 3 p: Summary of the longest series of pressure values south of 55°S: Monthly and annual average and extreme values of sea level pressure (mb) at ORCADAS, 1904 - 1981. Data of SIGNY ISLAND, 48 km due west of ORCADAS have been substituted for missing values of the last months of 1975. The values for the years 1903 - 1950 have been taken from Prohaska (1951), adding gravity correction and reduction to sea level. These data differ slightly from, and are preferable to, the values published in the World Weather Records. To every pressure value in this table, 900 mb have to be added.

	J	F	M	A	M	J	J	A	S	O	N	D	Year
a) monthly and annual averages, and standard deviations													
p	91.4	91.1	91.0	91.1	92.9	95.0	95.1	94.8	93.3	91.8	89.2	92.3	92.4
σ	4.1	3.9	4.0	4.0	5.4	5.0	4.7	4.6	4.9	4.5	5.3	4.5	1.5
b) maximum monthly and annual mean and year of occurrence													
p_x	102.8	103.6	101.4	100.9	109.2	105.5	109.5	107.5	106.2	102.2	103.1	104.6	96.2
y	1926	26	46	56	11	74	64	07	17	23	04	44	26
c) minimum monthly and annual mean and year of occurrence													
p_n	81.7	83.2	80.6	81.0	82.8	84.4	84.6	83.1	85.4	80.8	77.0	80.8	89.0
y	1959	48	35	73	61	05	76	76	05	14	60	52	59
d) amplitude b) - c)													
Δp	21.1	20.4	20.8	19.9	26.4	21.1	24.9	24.4	20.8	21.4	26.1	23.8	7.2
e) absolute extreme values; these data refer only to the time April 1903 to November 1956													
p_x	118.5	120.5	124.1	132.1	130.1	133.0	132.3	129.5	130.7	130.9	120.8	123.6	133.0
p_n	51.7	44.1	50.8	30.4	42.7	38.5	41.2	40.9	37.5	44.4	45.3	53.1	30.4
Δp	66.8	76.4	73.3	101.7	87.4	94.5	91.1	88.6	93.2	86.5	75.5	70.5	102.6

App. 4.[*] The modified Ekman solution applied to the inversion wind

The so-called Ekman solution describing the wind change with height in the boundary-layer, was developed by Taylor (1916) from Ekman's (1902) original application to the similar problem of ocean currents. Much later the atmospheric model has been modified to incorporate the sloped inversion thermal wind resulting from an exponentially shaped temperature profile over the sloping Antarctic plateau (Mahrt, 1969; Mahrt and Schwerdtfeger, 1970). The equations of this steady motion can be written:

$$0 = - g \frac{\Delta T}{T} \frac{\partial h}{\partial x} e^{-z/H} - \alpha \frac{\partial p}{\partial x} + fv + K \frac{\partial^2 u}{\partial z^2} \qquad (1)$$

$$0 = - g \frac{\Delta T}{T} \frac{\partial h}{\partial y} e^{-z/H} - \alpha \frac{\partial p}{\partial y} - fu + K \frac{\partial^2 v}{\partial z^2} \qquad (2)$$

Meaning of the right hand terms:
First : pressure gradient force resulting from the sloped inversion;
         h = height of the surface, H = mean thickness of inversion-layer,
Second: pressure gradient force above the inversion layer,
         i.e., the synoptic pressure force;
Third : the Coriolis forces;
Fourth: the friction forces.

Since the boundary-layer thermal wind components are $u_T = \frac{g}{f} \frac{\Delta T}{T} \frac{\partial h}{\partial y}$ and $v_T = \frac{g}{f} \frac{\Delta T}{T} \frac{\partial h}{\partial x}$, and the "synoptic" pressure force components

$$- \alpha \frac{\partial p}{\partial x} = - f v_g \quad \text{and} \quad - \alpha \frac{\partial p}{\partial y} = f u_g ,$$

the above equations become

$$0 = f v_T e^{- z/H} - f v_g + fv + K \frac{\partial^2 u}{\partial z^2} \qquad (3)$$

$$0 = f u_T e^{- z/H} + f u_g - fu + K \frac{\partial^2 u}{\partial z^2} \qquad (4)$$

---

[*]Appendices 4 and 5 were contributed by Dr. T. Parish.

Multiplying (4) by i and adding to (3) yields:

$$\frac{\partial}{\partial z} (u + iv) - \frac{if}{K} (u - u_g + iv - iv_g) = \frac{(if\ u_T - f\ v_T)}{K} e^{-z/H} \tag{5}$$

Since $\frac{\partial^2}{\partial z^2} (- u_g - i\ v_g) = 0$, we can rewrite (5):

$$\frac{\partial^2 Q}{\partial z^2} + \frac{i\ |f|}{K}\ Q = \frac{(if\ u_T - f\ v_T)}{K} e^{-z/H} \tag{6}$$

where $Q = (u - u_g - iv - iv_g)$. This differential equation has a general solution of:

$$Q = c_1 \exp\left(-i\ (\frac{if}{K})^{1/2} z\right) + c_2 \exp\left(i\ (\frac{if}{K})^{1/2} z\right) \quad .$$

As a first attempt at determining the particular solution, it is assumed to be of a form

$$Q = A \exp(-z/H).$$

Substituting this form into (6),

$$A = - \frac{(i|f|\ u_T - |f|\ v_T)}{(K + i\ H^2 |f|)}\ H^2$$

Thus, the general solution is:

$$Q = c_1 \exp\left(-i\ (\frac{if}{K})^{1/2} z\right) + c_2 \exp\left(i\ (\frac{if}{K})^{1/2} z\right) + A \exp(-z/H) \tag{7}$$

By specifying the proper boundary conditions:

at $z = 0$; $u = 0$, $v = 0$

at $z = \infty$; $u = u_g$, $v = v_g$,

eliminating the non-physical solution and separating the real components from the imaginary components, the modified Ekman solution of winds in the antarctic plateau boundary layer can be determined:

$$u = (B + v_g) \exp(-D) \sin (D)$$
$$- (C + u_g) \exp(-D) \cos (D)$$
$$+ C \exp(-z/H) + u_g$$

$$v = - (B + v_g) \exp(-D) \cos (D)$$
$$- (C + u_g) \exp(-D) \sin (D)$$
$$+ B \exp(-z/H) + v_g$$

where

$$B = - \frac{(H^4 |f|^2 v_T + H^2 |f| K u_T)}{(K^2 + H^4 |f|^2)}$$

$$C = \frac{H^2 |f| K v_T - H^4 |f|^2 u_T}{(K^2 + H^4 |f|^2)} \quad ,$$

$$D = (\frac{|f|}{2K} z)^{1/2}$$

App. 5. Formulation of the inversion wind,
used for the two-layer model in Section 3.1.3.

In this appendix the equations of motion describing the steady state
conditions of the boundary-layer over the gently sloping Antarctic plateau will
again be used to derive a formula for inversion winds. Only this time the
assumption is made that the inversion-layer, colder than the free atmosphere
above it, has constant density.

The horizontal pressure gradient force, $\vec{F}$, shall be considered as the
vector sum of the terrain-induced horizontal pressure gradient force resulting
from the existence of an inversion over sloping terrain and the synoptic-scale
pressure gradient force above the inversion. Mathematically, this can be
expressed:

$$\vec{F} = - \alpha \frac{\partial p}{\partial x} \vec{i} - g \frac{\Delta T}{T} \frac{\partial h}{\partial x} \vec{i} - \alpha \frac{\partial p}{\partial y} \vec{j} - g \frac{\Delta T}{T} \frac{\partial h}{\partial y} \vec{j} \tag{1}$$

By rotating the horizontal coordinates such that $\vec{F}$ is directed along one axis,
the scalar equations of motion for the inversion wind can be written:

$$0 = F + fv - \frac{k}{H} Vu \tag{2}$$

$$0 = - fu - \frac{k}{H} Vv . \tag{3}$$

Here, V is the magnitude of the wind and can be written $V = (u^2 + v^2)^{1/2}$; the
scalar components u and v are expressed as follows:

$u = V \cos\gamma , \quad v = V \sin\gamma$

where γ is the deviation angle of the wind from the axis of the pressure gradient force. The last terms in each expression represent the friction forces,
with friction being proportional to the square of the wind speed and inversely
proportional to the inversion layer thickness, H.

Solving for V is accomplished by first eliminating the Coriolis force:

$$0 = Fu + fuv - \frac{k}{H} \, Vu^2$$

$$+ \quad 0 = \quad - fuv - \frac{k}{H} \, Vv^2$$

$$\overline{\qquad\qquad\qquad\qquad\qquad\qquad}$$

$$0 = Fu - \frac{k}{H} \, (u^2 + v^2) \tag{4}$$

Noting that $V^2 = u^2 + v^2$, (4) can be rewritten

$$0 = FV \cos\gamma - \frac{k}{H} \, V^3$$

or,

$$V = (\frac{HF}{k} \cos\gamma)^{1/2} \quad . \tag{5}$$

Solving for V can also be accomplished by eliminating the friction terms:

$$0 = Fv + fv^2 - \frac{k \, Vuv}{H}$$

$$- \, 0 = \quad - fu^2 - \frac{k \, Vvu}{H}$$

$$\overline{\qquad\qquad\qquad\qquad\qquad\qquad}$$

$$0 = Fv + f(u^2 + v^2) \tag{6}$$

Rearranging (6) using the above expressions for the wind components,

$$0 = FV \sin\gamma + f \, V^2$$

or

$$V = - \frac{F \sin\gamma}{f} \quad . \tag{7}$$

To determine an analytic solution for the deviation angle γ, note that

$$V^2 = \frac{H}{k} F \cos\gamma = (- \frac{F \sin\gamma}{F})^2 \quad .$$

Since $\sin^2 \gamma = 1 - \cos^2 \gamma$,

$$\frac{HF}{k} \cos\gamma = \frac{F^2}{f^2} (1 - \cos^2\gamma)$$

or

$$\cos^2\gamma + \frac{Hf^2}{Fk} \cos\gamma - 1 = 0 \qquad\qquad (8)$$

This is a quadratic expression, the roots of which can be written

$$\cos\gamma = - \frac{\frac{Hf^2}{Fk} \pm [(\frac{Hf^2}{Fk})^2 + 4]^{1/2}}{2}$$

The deviation angle of the wind from the fall line can be easily computed by noting the difference of the pressure gradient force, F , from the fall line.

REFERENCES

Abbreviations used

A.J.U.S.   Antarctic Journal of the United States
B.A.M.S.   Bulletin, American Meteorological Society
B.L.M.   Boundary Layer Meteorology
J.A.M.   Journal of Applied Meteorology
J.A.S.   Journal of Atmospheric Science
J. Clim.   Journal of Climatology
J.G.R.   Journal of Geophysical Research
J. of Glac.   Journal of Glaciology
M.W.R.   Monthly Weather Review
S.A.E.I.B.   Soviet Antarctic Expeditions Information Bulletin

Ackley, S.F., 1979a. Mass-balance aspects of Weddell Sea pack ice. J. of Glac.,
     24(90): 391-405.
Ackley, S.F., 1979b. A review of sea-ice weather relationships in the southern
     hemisphere, and: Sea-ice atmosphere interactions in the Weddell Sea using
     drifting buoys. Proceedings of the Canberra Symposium on Sea Level, Ice,
     and Climatic Change, I A H S Publ. Nr. 131: 127-159 and 177-191.
Adachi, T., 1979. Numerical simulation of katabatic wind profiles at Syowa
     Station, Antarctica. Antarctic Record (Japan), 67: 64-74.
Allison, I., and Morrissy, J.V., 1981. Automatic meteorological stations and
     data buoys in the Antarctic - recent Australian experience. Preprint volume
     of papers presented at the University of Melbourne, May 1981: 13 pp.
Alvarez, J.A., and Lieske, B.J., 1959. The Little America Blizzard of May 1957.
     In: Antarctic Meteorology, Proceedings of the Symposium held in Melbourne,
     February 1959. Pergamon Press 1960: 115-127.
American Polar Society, 1963. Third Little America is afloat 300 miles from
     original site. The Polar Times, 56: 3.
Amundsen, R., 1912. The South Pole. Translation. A.G. Chater, London; two vol.
ANARE DATA REPORTS, 1956-68. Austral. Nat. Antarctic Res. Exp. Series D,
     Meteorology. Annual volumes prepared by the Bureau of Meteorology, Melbourne.
Anderson, T., 1981. Bergeron and the Oreigenic Maxima of Precipitation. Tor
     Bergeron memorial volume, Pure and Applied Geophysics, 119(3): 558-576.
Anonymous, 1977. Cold breaks records at South Pole. A.J.U.S. 12(1 + 2): 51.
Anonymous, 1978. Weather from the ends of the Earth. MOSAIC, Bimonthly journal
     of the National Science Foundation, Washington, D.C., 9(5): 38-46.
Anthes, R.A., Panofsky, H.A., Cahir, J.J., and Rango, A., 1975. The Atmosphere,
     p. 264. C.E. Merrill Publ. Cy.: 339 pp.
Averyanov, B.G., 1972. The climate of Vostok. Scientific Results of the USSR
     Antarctic Expeditions, volume 60: 175-226. (In Russian).
Averyanov, B.G., 1975. The climate of Molodezhnaya 1963-72. Scientific Results
     of the USSR Antarctic Expeditions, vol. 65: 95-126. (In Russian).

Baker, P.E., 1968. Investigations of the 1967 and 1969 Volcanic Eruptions on
     Deception Island, South Shetland Islands. Polar Record, 14 (93): 823-827.
Baker, P.E., Davies, T.G., and Roobol, M.J., 1969. Volcanic activity at
     Deception Island, 1967 and 1969. Nature, London, 224 (5219): 553-560.
Ball, F.K., 1960. Winds on the Ice Slopes of Antarctica. Antarctic Meteorology,
     Proceedings of the Symposium held at Melbourne 1959. Pergamon Press, Oxford:
     9-16.
Barkow, E., 1913. Vorläufiger Bericht über die meteorologischen Beobachtungen
     der Deutschen Antarktischen Expedition 1911-1912. Veröff. Preuss. Met.
     Institut 4(11): 11 pp.
Barkow, E., 1924. Die Ergebnisse der meteorologischen Beobachtungen der
     Deutschen Antarktischen Expedition 1911-1912. Veröff. Preuss. Met. Institut
     7 (6): 166 pp.

Barrie, J.M., 1913. Scott's Last Expedition. John Murray, London, reprinted 1954: 521 pp.

Barry, R.G., 1981. Mountain Weather and Climate. Methuen Publ. London and New York: 313 pp.

Bassett, W.E., 1923. Results of tests made on a Robinson cup anemometer for Sir Douglas Mawson. Australasian Antarctic Expedition 1911-1914, Scientific Reports Series B, Vol. IV, Appendix A: 3-4.

Benson, C.S., 1971. Stratigraphic studies in the snow at Byrd Station, Antarctica. A.G.U. Antarctic Research Series 16: 333-353.

Bergeron, T., 1960. Operation and results of "Project Pluvius". Monogr. 5, Am. Geophys. Union: 152-157.

Boddington, J., editor, 1980. 1910-1916 Antarctic Photographs by H. Ponting and F. Hurley. Foreward by Vivian Fuchs. St. Martin's Press New York, 10010, 120 pp.

Bodman, G., 1908-1910. Meteorologische Ergebnisse der Schwedischen Suedpolar-Expedition. in: Wissenschaftl. Erg. der Schwedischen Suedpolar-Expedition 1901-03, Stockholm, Volume II, 282 pp.

Borchgrevink, C.E., 1901. First on the Antarctic continent; being an account of the British Antarctic Expedition 1898-1900. London,333 pp.

Boujon, H., 1954. Les observations météorologiques de Port-Martin en Terre Adélie. Fasc. I, Conditions atmosphériques en surface et en altitude du 14. II. 1950 au 14. I. 1951. Relevés quotidiens et commentaires. Expéd. Pol. Franc., Résultats Scientif., S 5: 167 pp.

Bromwich, D.H., 1978. Some considerations in deriving poleward water vapor transport values for coastal East Antarctica. A.J.U.S. 13(4): 196-197.

Bromwich, D.H., 1979. Precipitation and accumulation estimates for East Antarctica, derived from rawinsonde information. Research Report, Dept. of Meteorology, Univ. of Wisconsin: 142 pp.

Bromwich, D.H., and Kurtz, D.D., 1982. Experiences of Scott's Northern Party: Evidence fore a relationship between winter katabatic winds and the TERRA NOVA BAY polynya. Polar Record 21(131): 137-146.

Brost, R.A., and Wyngaard, J.C., 1978. A model study of the stable stratified planetary boundary-layer. J.A.S., 35: 1427-1440.

Brown, C.W., and Keeling, Ch. D., 1965. The concentration of atmospheric carbon dioxide in Antarctica. J.G.R., 70: 6077-6085.

Budd, W.F., 1975. Antarctic mass balance studies. In: Progress in Australian Hydrology 1965-1974. Austral Nat. Comm. to UNESCO Austral. Gvt. Publ. Service Canberra: 88-95.

Budd, W.F., Dingle, W., and Radok, U., 1966. The Byrd snowdrift project; outline and basic results. In: Studies in Antarctic meteorology, M. Rubin (ed.). Am. Geophys. Union, Antarctic Res. Ser. 9: 71-134.

Budd, W., Landon Smith, I., and Wishart, E., 1967. "The Amery Ice Shelf". In: Physics of snow and ice. Proceedings of the Conference on Low Temperature Science, Vol. 1, Part 1 (quoted from Kotlyakov, 1977).

Bull, C., 1971. Snow accumulation in Antarctica. In: Research in the Antarctic, L.O. Quam, editor, Am. Ass. Adv. Sci.: 367-421.

Burdecki, F., 1969, 1970. The Climate of SANAE. Part I: NOTOS 18, 1969: 65, Part II: NOTOS 19, 1970: 62, published by the Weather Bureau of South Africa.

Burman, E.A., and Yemshanova, N.V., 1973. Some characteristics of the vertical structure of strong drainage winds in East-Antarctica. S.A.E.I.B. 87, Engl. translation VIII (9): 492-494.

Burrows, C.J., 1976. Icebergs in the Southern Ocean. New Zealand Geographer 32: 127-138.

Carleton, A.M., 1981. Monthly variability of satellite-derived cyclonic activity for the Southern Hemisphere winter. J. Climat., 1:21-38.

Carroll, J.J., 1982. Long-term means and short-time variability of the surface energy balance components at the South Pole. JGR 87 (C6): 4277-4286.

244

Carroll, J.J., and Fitch, B.W., 1981. Effects of solar elevation and cloudiness on snow albedo at the South Pole. JGR, 86 (C6): 5271-5276.

Cherry-Garrard, A., 1923. The worst journey in the World. Reprinted in Penguin Books, 1970: 401 pp.

Clausen, H.B., Dansgaard, W., Nielsen J.O., and Clough, J.W., 1979. Surface Accumulation on Ross Ice Shelf. A.J.U.S. XIV (5): 68-72.

Clinch, N.B., 1967. First conquest of Antarctica's highest peaks. National Geographic Magazine 131 (6): 836-863.

Court, A., 1948. Wind Chill. B.A.M.S. 29(10): 487-493.

Court, A., 1949. Meteorological Data for Little America III. M.W.R. Suppl. No. 48. U.S. Govt. Printing Office, Washington, D.C., 150 pp.

Colvill, A.J., 1977. Movement of Antarctic ice fronts measured from satellite imagery. Polar Record 18 (115): 390-394.

Curtis, R.H., 1908. Discussion of the Observations..., chapter XII. In: National Antarctic Expedition 1901-04, Meteorology, Part 1, published by the Royal Society, London: 383-415.

Dalrymple, P., 1966. A physical climatology of the Antarctic Plateau. In: Studies in Antarctic Meteorology, M.J. Rubin, editor, A.G.U. Antarctic Research Series 9: 195-231.

Dalrymple, P., Lettau, H.H., and Wollaston, S., 1966: South Pole micrometeorology program, data analysis. In: Studies in Antarctic Meteorology (M.J. Rubin, editor), Am. Geophys. Union Antarctic Research Series 9: 13-58.

Dalrymple, P.C., and Frostman, T.O., 1971. Some aspects of the climate of interior Antarctica. In: Research in the Antarctic, AAAS-Publication No. 93: 429-442.

Dalrymple, P.C., and Stroschein, L.A., 1977. Micrometeorological System: Installation, Performance, and Problems. Meteorological Studies at Plateau Station, Antarctica. AGU Antarctic Research Series, 25, Paper 1, 39 pp.

Damon, P.E., and Kunen, S.M., 1976. Global cooling? No, southern hemisphere warming trends may indicate the onset of the CO_2 "greenhouse" effect. SCIENCE 193 (4252): 447-453.

Dare, P.M., 1981. A study of the severity of the midwestern winters of 1977 and 1978 using heating degree days determined from both measured and windchill temperatures. B.A.M.S. 62(7): 974-982.

Defant, A., 1933. Der Abfluss schwerer Luftmassen auf geneigtem Boden, nebst einigen Bemerkungen zu der Theorie stationaerer Luftstroeme. Sitz. Ber. Preuss. Akad. d. Wiss., Berlin, 18: 624-635.

Defant, F., 1951. Local winds. In Compendium of Meteorology (T.F. Malone, editor), Am. Met. Soc., Boston, Mass., 655-672.

De La Canal, L.M., 1963. Bases para el pronóstico del estado de los hielos de mar en algunas regiones del sector antártico Argentino. Publicación H 412 del Servicio de Hidrografia Naval, Secretaría de Marina, Rep. Argentina, 65 p.

Dolgina, I.M., Marshunova, M.A., and Petrova, L.S., 1976 and 1977. Arctic and Antarctic Scientif. Res. Institute, Leningrad. Reference Book of the Climate Antarctica's (in Russian), Vol. I. Radiation: 213 pp., Vol. II. Surface Data: 493 pp.

Dorsey, H.G., 1945. An Antarctic Mountain Weather Station. Proceedings of the Am. Phil. Soc. Philadelphia. 89(1): 344-363.

Du Faur, E., 1907. The effect of polar ice on the weather. Journal and Proceedings of the Royal Society of New South Wales, 41: 176-189.

Ekman, V.W., 1902. On the influence of the Earth's rotation on ocean-currents. Arkiv for Matematik, Astronomi och Fysik, vol. II (11).

Elsaesser, H.W., 1983. The climatic effect of CO_2: a different view. Paper presented at the Second Conference on Climate Variations, New Orleans, La., January 1983.

ESSA, U.S. Dept. of Commerce, 1968 and 1970. Climatological Data for Antarctic
    Stations, Jan.-Dec. 1966, #9: 70-206; Jan. 1967-Dec. 1968, #10: 205-246.

Falconer, R., 1968. Windchill, a useful wintertime weather variable.
    Weatherwise 21(6): 227-229, 255.
Farman, J.C., and Hamilton, R.A., 1978. Measurements of radiation at the
    Argentine Islands and Halley Bay, 1963-72. British Antarctic Survey,
    Scientific Reports 99: 53 pp.
FGGE, 1981. The Global Weather Experiment, daily global analyses, Part 3,
    June-August 1979. European Center for Medium Range Weather Forecasts,
    Reading, UK, unpaged.
Ficker, H.v., 1906. Der Transport kalter Luftmassen ueber die Zentral-Alpen.
    Denkschriften der Math.-Naturwiss. Klasse der K. Akad. der Wiss., Wien,
    Band 80: 70 pp.
Fleming, R.J., Kaneshige, T.M., McGovern, W.E., and Bryan, T.E., 1979. The
    Global Weather Experiment II. The second special observing period. B.A.M.S.
    60(11): 1316-1322.
Fletcher, J.O., 1969. Ice Extent on the Southern Ocean and its Relation to
    World Climate. Memorandum RM-5793-NSF. The Rand Corporation, California,
    108 p.
Fletcher, J.O., Radok, U., and Slutz, R., 1982. Climatic signals of the
    Antarctic Ocean. JGR 87 (C6): 4269-4276.
Flohn, H., 1978. Comparison of antarctic and arctic climate and its relevance
    to climate evolution. ICSU, SCAR, Proc. Symp. August 1977: 3-13.
Flohn, H., 1980. Possible climatic consequences of a man-made global warming.
    IIASA RB-80-30, Laxenburg, Austria: 81 pp.
Frostman, T.O., Martin, D.W., and Schwerdtfeger, W., 1967. Annual and semi-
    annual variations in the length of day, related to geophysical effects.
    JGR 72(20): 5065-5073.
Fujii, Y., 1979. Sublimation and condensation at the ice sheet surface of
    Mizuho Station, Antarctica. Antarctic Record (Japan), 67: 51-63.
Fujii, Y., and Kusunoki, K., 1982. The role of sublimation and condensation
    in the formation of the ice sheet surface at Mizuho Station, Antarctica.
    JGR 87(C6): 4293-4300.

Geophysical Monitoring for Climatic Change, 1981 and 1982. No. 9, Summary
    Report 1980, No. 10, Summary Report 1981. NOAA, Env. Res. Lab. Boulder:
    163 pp.
Giovinetto, M.B., 1968. Glacier landforms of the Antarctic coast, and the
    regimen of the Inland Ice. Ph.D. thesis, University of Wisconsin at
    Madison. 164 pp.
Gloersen, P., Wilheit, T.T., Chang, T.C., Nordberg, W., and Campbell, W.J.,
    1973. Microwave maps of the polar ice of the Earth. Preprint X-652-73-269,
    Goddard Space Flight Center, Greenbelt, Md.; 37 p., 7 plates.
Gonzalez-Ferran, O., 1982. The Seal Nunataks: an active volcanic group on the
    Larsen Ice Shelf, West Antarctica. Paper presented at the SCAR/IUGS
    Symposium on Antarctic Earth Sciences, University of Adelaide, South
    Australia.
Gordon, A.L., and Taylor, H.W., 1975. Seasonal Change of Antarctic Sea Ice
    Cover. Science 187 (4174): 346-47.
Gordon, A.L., and Molinelli, E.J., 1982. Southern Ocean Atlas, Part:
    Thermohaline and chemical distributions and the Atlas data set. Columbia
    University Press, New York.
Grandoso, H.N., and Nuñez, J.E., 1955. Analisis de una Situatión de Bloqueo
    en la parte Austral de América del Sur. Meteoros, Buenos Aires, 5(1): 35-54.
Griffiths, J.F. and Soliman, K.H., 1972. The Northern Desert Sahara. Chapter 3
    of Climates of Africa. J.F. Griffiths, editor. H.E. Landsberg, editor-in-
    chief, ELSEVIER SCI PUBL. CO. Amsterdam: 75-132.

Guymer, L.B., and Le Marshall, J.F., 1981. Impact of FGGE buoy data on Southern Hemisphere analyses. B.A.M.S., 62(1): 38-47. also in: Austral. Met. Mag. 1980, 28(1): 19-42.

Hall, F.F., 1978. Boundary-layer climatologies from acoustic sounder investigations. A.M.S. Preprint Volume, Fourth Symposium on met. obs. and instrumentation, April 1978: 330-332.

Hann, J. v., 1909. Die meteorologischen Ergebnisse der englischen antarktischen Expedition, 1901-04. Meteorolog. Zschr. 26: 289-301.

Hann, J. von, 1932. Handbuch der Klimatologie. Fourth edition, amplified by K. Knoch. 444 pp.

Hare, F.K., and Hay, J.E., 1974. The climate of Canada and Alaska. Chapter 2 in volume 11 of the World Survey of Climatology, Climates of North America, edited by R.A. Bryson and F.K. Hare: 49-188.

Heap, J.A., 1964. Pack Ice. Chapter XIX in: Antarctic Research, a Review of British Scientific Achievement in Antarctica. Edited by R.E. Priestley et al., London, Butterworth, pp. 308-317.

Helbig, A., 1979. Meteorologische Prozesse in der Grundschicht ueber der Antarktis. Zschr. f. Met., 29(2): 111-122.

Hisdal, V., 1958. Surface observations, Part 2: Wind. Norw.-Brit.-Swed. Antarctic Exp. 1949-1952, Scientific Results, Vol. 1: 67-121. Published by the Norsk Polarinstitutt.

Hogan, A., Barnard, S., Samson, J., and Winters, W., 1982. The transport of heat, water vapor, and particulate material to the South Polar Plateau. J.G.R. 87 (C6): 4287-4292.

Hollin, J.T., 1972. Interglacial climates and antarctic ice surges. Quart. Res. 2: 401-408.

Hoinkes, H.C., 1961. Studies of glacial meteorology at Little America V, Antarctica. Proceedings of the Symposium on Antarctic Glaciology, Int. Ass. of Sci. Hydrology, Publ.55: 29-48.

Huschke, R.E. (editor), 1959. Glossary of Meteorology. American Meteorological Society, Boston, Mass.: 638 p.

Ishikawa, N., Kobayashi, S., Ohata, T., and Kawaguchi, S., 1982. Radiation data at Mizuho station, Antarctica, in 1980. JARE Reports 73 (Met. 11): 195 pp.

Jacka, T.H., 1981. Antarctic temperature and sea-ice extent studies. Antarctica: Weather and Climate, Preprint volume University of Melbourne, May 1981: 10 pp.

Japan Meteorological Agency, 1964-1983. Antarctic Meteorological Data at Syowa Station, vol. 2-22.

JARE DATE REPORTS, 1974-1983. Japanese Antarctic Research Expeditions. National Institute of Polar Research, Tokyo. Meteorology, issues 1-13; Glaciology 1-8.

Jenne, R.L., Crutcher, H.L., van Loon, H., and Taljaard, J.J., 1971. Climate of the Upper Air, Southern Hemisphere, Vol. III: Vector Mean Geostrophic Winds. NAVAIR 50-IC-57, NCAR TN-STR-58: 60 maps.

Jouzel, J., Merlivat, L., Petit, J.R., and Lorius, C., 1983. Climatic information over the last century deduced from a detailed isotop record in the South Pole snow. JGR 88 (C4): 2693-2703.

Keeling, Ch. D., 1960. The concentration and isotopic abaundances of carbon dioxide in the atmosphere. TELLUS, 12: 200-203.

Keeling, C.D., Adams, J., Ekdal, C., and Guenther, P., 1976. Atmospheric carbon dioxide variations at the South Pole. TELLUS 28: 552-564.

Kidson, E., 1946. Discussion of observations at Adelie Land. Australasian Antarctic Expedition 1911-1914, Sci. Rept., Ser. B, 6: 121 pp.

Knittel, J., 1976. Ein Beitrag zur Klimatologie der Stratosphaere der
    Suedhalbkugel. Met. Abh. F.U. Berlin, Neue Folge, 2(1): 294 pp.
Kobayashi, S., 1978. Vertical structure of katabatic winds at Mizuho Plateau.
    Memoirs of the National Institute of Polar Research Tokyo, Special Issue
    No. 7: 72-80.
Kobayashi, S., and Ishida, T., 1979. Some characteristics of turbulence in
    katabatic winds over MIZUHO Plateau, East Antarctica. Antarctic Record,
    Japan, 67: 75-85.
Koerner, R.M., 1971. A stratigraphic method of determining the snow accumula-
    tion rate at Plateau Station, Antarctica, and application to South Pole -
    Queen Mary Land traverse. In: Antarctic Snow and Ice Studies II, AGU
    Antarctic Research Series 16: 225-238.
Kotlyakov, V.M., Losev, K.S., and Loseva, I.A., 1978. The ice budget of
    Antarctica. From Izvestiya Akademii Nank SSSR, 1977, translation in
    Polar Geography 2 (4): 251-262.
Kozo, T.L., 1980. Mountain barrier baroclinity effects on surface winds along
    the Alaskan Arctic Coast. Geophys. Res. Let., 7 (5): 377-380.
Kuhn, M., 1969. Preliminary Report on Meteorological Studies at Plateau
    Station, Antarctica. Published by the Meteorology Department, University
    of Melbourne. No number. 20 pp.
Kuhn, M., 1970. "Ice crystals and solar halo displays, Plateau Station, 1967".
    International Symposium on Antarctic Glaciological Exploration (I. Ass. Sci.
    Hydr.): 298-303.
Kuhn, M., 1977. Precipitation of whisker crystals at surface temperatures below
    -60°C on the East-antarctic Plateau. Collection of Contributions at CPM
    Sessions Joint IAGA/IAMAP Assembly, Seattle, Washington, 1977: 60-66.
Kuhn, M., 1978. Optical phenomena in the Antarctic atmsophere. Paper 9 in
    Meteorological Studies at Plateau Station, A.G.U. Antarctic Research Series
    Vol. 25 (J.A. Businger, editor): 129-155.
Kuhn, M., 1980. Antarktic - die grösste Wüste der Welt. UMSCHAU 80 (22):
    675-681.
Kuhn, M., Kundla, L.S., and Stroschein, L.A., 1977. The radiation budget at
    Plateau Station, Antarctica, 1966-67. Paper 5 in Meteorological Studies
    at Plateau Station. A.G.U. Antarctic Research Series, Vol. 25.
    (J.A. Businger, editor): 41-73.
Kuhn, M., Lettau, H. and Riordan, A., 1977. Stability-related wind spiraling
    in the lowest 32 meters. Paper 7 in Meteorological Studies at Plateau
    Station, A.G.U. Antarctic Research Series, Vol. 25, (A.J. Businger, editor):
    93-111.
Kuhn, M., and Siogas, L., 1978. Spectroscopic studies at McMurdo, South Pole,
    and Siple stations during the austral summer 1977-78. A.J.U.S. 13 (4):
    178-179.
Kuhn, P.M., and Weickmann, H.K., 1969. High altitude radiometric measurements
    of cirrus. J.A.M. 8 (2): 147-154.
Kutzbach, G., and Schwerdtfeger, W., 1967. Temperature variations and vertical
    motion in the free atmosphere over Antarctica in the winter. WMO Technical
    Note 87: 225-248.
Kyle, T.H., and Schwerdtfeger, W., 1974. Synoptic scale wind effects on the
    ice cover in the western Weddell Sea. A.J.U.S., 9 (5): 212-213.

La Grange, J.J., 1963. Trans-Antarctic Expedition 1955-58, Meteorology 1,
    Shackleton, South Ice, and the Journey across Antarctica. Trans-Antarctic
    Expedition 1955-58. Scientific Reports No. 13, London. 93 pp.
Lamb, H.H., 1967. On climatic variations affecting the far south. Proc. Symp.
    Polar Meteorology, Geneva 1966, WMO Technical Note, 87, 428-453.
Lax, J.N., and Schwerdtfeger, W., 1976. Terrain-induced vertical motion and
    occurrence of ice crystal fall at South Pole in summer. A.J.U.S. 11 (3):
    144-145.

248

Learmont, J.S., 1950. Master in Sail, London, Percival Marshall Publ., on p. 120 quotation from an article by Mr. Huycke in the journal 'Sea Breezes', October 1947.

Le Marshall, J.F., and Kelly, G.A.M., 1981. A January and July climatology of the Southern Hemisphere based on daily numerical analyses 1973-1977. Austral. Met. Mag., 29 (3): 115-123.

Le Quinio, R., 1956. Expéditions Polaires Francaises, Resultats Scientifiques, #S.V. Les observations météorologiques de Port Martin, en Terre Adélie. Fasc. IV, Conditions atmospheriques en surface et en altitude, 1950-1952. 97 pp.

Lettau, H.H., 1966. A case study of katabatic flow on the South Polar Plateau. In: Studies in Antarctic Meteorology, AGU Antarctic Research Series, 9: 1-12.

Lettau, H., 1967. Small to Large-Scale Features of Boundary Layer Structure over Mountain Slopes. In: Proceedings of the Symposium on Mountain Meteorology, June 1967, Fort Collins, Colorado, (E.R. Reiter and J.L. Rasmussen, editors): 1-74.

Lettau, H.H., 1971. Antarctic atmosphere as a test tube for meteorological theories. In: Research in the Antarctic (L.O. Quam, editor), papers presented at the Dallas Meeting 1968 of the AAAS. Publ. 93: 429-475.

Lettau, H., 1977. Climatonomic Modeling of Temperature Response to Dust Contamination of Antarctic Snow Surfaces. B.L.M. 12 (2): 213-229.

Lettau, H., 1978. Exploring the World's Driest Climate. Chapter XII of Scientific Results of University of Wisconsin Field Studies during July 1964 in the Peruvian Desert, edited by H. and K. Lettau. IES Report 101, Center for Climatic Research, Institute for Environmental Studies, Univ. of Wisconsin: 182-248.

Lettau, H.H., and Schwerdtfeger, W., 1967. Dynamics of the surface-wind regime over the interior of Antarctica. A.J.U.S., 2 (5): 155-158.

Lettau, H., Riordan, A., and Kuhn, M., 1977. Air temperature and two-dimensional wind profiles in the lowest 32 meters as a function of bulk stability. Paper 6 in Meteorological Research Series, Vol. 25 (A.J. Businger, editor): 77-91.

Lewis, D., 1975. Ice Bird, The First Single-handed Voyage to Antarctica. W.W. Norton & Co., New York, 224 pp; permission to copy the last ten lines of page 179 granted.

Liljequist, G.H., 1956 and 1957. Energy exchange of an antarctic snow-field. Norw.-Brit.-Swed. Antarctic Expedition 1949-1952, Scientific Results, Part 1, vol. 2: 4 issues, 289 pp.

Limbert, D.W.S., 1974. Variations of the mean annual temperature for the Antarctic Peninsula, 1904-72. Polar Record 17 (108): 303-306.

Lindzen, R.S., Hou, A.Y., and Farrell, B.F., 1982. The role of convective model choice in calculating the climate impact of doubling CO_2. J.A.S. 39 (6): 1189-1205.

Linkletter, G.O., and Warburton, J.A., 1976. A note on the contribution to the accumulation on the Ross Ice Shelf, Antarctica. J. of Glac. 17 (76): 351-354.

List, R.J., 1958. Smithsonian Meteorological Tables. Smiths. Misc. Coll., vol. 114: 527 pp.

Loewe, F., 1956. Etudes de Glaciologie en Terre Adélie, 1951-52. Expéditions Polaires Francaises, Paris (B. Valtat, editor), No. 1247: 159 pp.

Loewe, F., 1962. On the mass economy of the interior of the Antarctic ice cap. J.G.R. 67 (13): 5171-5177.

Loewe, F., 1967. The water budget in Antarctica. In: Proc. Symp. Pacific Sciences. JARE Sci. Reports, Spec. Issue 1: 101-110.

Loewe, F., 1970. Screen temperatures and 10 m temperatures. J. of Glac. 9 (56): 263-268.

Loewe, F., 1972. The Land of Storms. WEATHER, London, 27 (3): 110-121.

Loewe, F., 1974a. Die taegliche Windschwankung ueber dem Innern von Inlandeisen im Sommer. Arch. Met. Geoph. Biokl., Serie B, 22: 219-232.

Loewe, F., 1974b. Considerations concerning the winds of Adelie Land. Zeitschrift fuer Gletscherkunde und Glazialgeologie, 10: 189-197.

Lorius, C., 1983. Some data on climate, CO_2, aerosols and ice thickness changes from antarctic ice cores. Lecture given at the Workshop on potential CO_2-induced changes in the environment of West Antarctica Geophysical and Polar Research Center, U. of Wisconsin at Madison, July 1983.

Madigan, C.T., 1929. Meteorology of the Cape Denison station. Australasian Antarctic Expedition 1911-14, Scientific Reports, Serie B, 4: 286 pp.

Maenhout, W., Zoller, W.H., and Coles, D.G., 1979. Radionuclides in the South Polar Atmosphere. J.G.R. 84 (C6): 3131-3138.

Mahrt, L.J., 1969. The Wind Regime of the Sloped Inversion Friction Layer in the Interior of Antarctica. M.S. Thesis, University of Wisconsin at Madison: 65 p.

Mahrt, L.J., and Schwerdtfeger, W., 1970. Ekman spirals for exponential thermal wind. B.L.M., 1, 137-145.

Malberg, H., 1969. Der Einfluss der Gebirge auf die Luftdruckverteilung am Erdboden. Meteorolog. Abhandl. d. Institut f. Meteorologie und Geophysik d. Universität Berlin, 71 (2), 66 pp.

Marshunova, M.S., 1980. Short-wave radiation regime in Antarctica based on 20 years of observation. in "Climatic research in the Antarctic", edited by I.M. Dolgina, Leningrad, Gidrometeoizdat: 11-19. In Russian.

Martin, D.W., 1968a. Satellite studies of cyclonic developments over the Southern Ocean. Tech. Rept. 9, IAMRC, Bureau of Meteorology Melbourne, 64 pp., and Ph.D. thesis, Dept. of Meteorology, Univ. of Wisconsin: 144 pp.

Martin, D.W., 1968b. Cyclones over the Southern Oceans. A.J.U.S. 3(5): 191-192.

Mather, K.B., 1960. Katabatic winds south of Mawson. Antarctic Meteorology, 1959 Symposium Proceedings: 317-320.

Mather, K.B., 1962. Further observations on sastrugi, snow dunes and the pattern of surface winds in Antarctica. Polar Record 11: 158-171.

Mather, K.B., and Miller, G.S., 1967. Notes on Topographic Factors Affecting the Surface Wind in Antarctica. University of Alaska Technical Report UAG-R-189: 125 p.

Mawson, D., 1915. The home of the blizzard, being the story of the Australian Antarctic Expedition 1911-14. Heinemann, London, 2 volumes.

Mayes, P.R., 1981. Recent trends in antarctic temperature. Climate Monitor 10 (4): 96-100.

Mechoso, C.R., 1980. The atmospheric circulation around Antarctica: Linear stability and finite-amplitude interactions with migrating cyclones. J.A.S. 37 (10): 2209-2233.

Meinardus, W., 1928. Die Luftdruckverhältnisse und ihre Wandlungen südlich von 30°S Breite. Part 3, Vol. III of Deutsche Südpolar-Expedition, E.v. Drygalski, editor, Walter de Gruyter, publisher, Berlin and Leipzig 1928: 131-307.

Meinardus, W., 1938. Klimakunde der Antarktis. In: Koeppen und Geiger, Handbuch der Klimatologie, Band IV (U): 180 pp.

Meinardus, W., and Mecking, L., 1911. Tägliche synoptische Wetterkarten der höheren südlichen Breiten von Oktober 1901 bis März 1904. In: Deutsche Südpolar-Expedition 1901-1903, Meteorologischer Atlas, Berlin 1911.

Meinardus, W., and Mecking, L., 1915. Meteorologischer Atlas mit Monats - und Tageskarten vom 1. Okt. 1901 bis 31. März 1904 südlich von 30°S.

Mellor, M., 1977. Engineering properties of snow. J. of Glac. 19 (81): 15-66.

Miller, St. A., 1974. An analysis of heat and moisture budgets of the inversion layer over the Antarctic Plateau, for steady state conditions. Research Report, Department of Meteorology, University of Wisconsin: 68 p.

Miller, St. A., and Schwerdtfeger, W., 1972. Ice crystal formation and growth in the warm layer above the antarctic temperature inversion. A.J.U.S. 7 (5): 170-171.

Mitchell, W.F., Sallee, R.W., and Snell, A.W., 1970. Antarctic Forecasters Handbook. Antarctic Support Activities, Detachment Charlie. 200 pp.

Monographies de la Meteorologie Nationale, 1969. Données Climatiques de la station DUMONT D'URVILLE. M.M.N. Paris, 72: 63 pp.

Morgan, V.I., and Jacka, T.H., 1979. Mass balance studies in East Antarctica. In: Sea level, Ice, and Climatic Change, Proceedings of the Canberra Symposium, December 1979. IHAS Publ. 131: 253-260.

Mossman, R.C., 1907. Results of the Scottish National Antarctic Expedition 1902-04 (W.S. Bruce). Vol. II, Part 1, Meteorology.

Mullen, S.L., 1979. An investigation of small synoptic-scale cyclones in polar air streams. M.W.R. 107 (12): 1636-1647.

Munk, W.H., and MacDonald, G.J.F., 1960. The rotation of the Earth. The rotation of the Earth. Cambridge University Press, 323 pp.

Murcray, D.G., Brooks, J.N., Murcray, F.H., and Williams, W.J., 1974. 10 to 12 μm spectral emissivity of a cirrus cloud. J.A.S. 31 (10): 1940-1942.

Nansen, F., 1897 and 1898. In Nacht und Eis. Die Norwegische Polarexpedition 1893-1896. Brockhaus, Leipzig, Vol. 1: 527 pp., Vol. 2: 539 pp.

NASA, National Aeronautics and Space Administration, 1971. The Best of Nimbus. NASA Publ. 9 G45-80: 119 pp.

National Gallery of Victoria, Melbourne, 1979. 1910-1916 Antarctic Photographs taken by H. Ponting and F. Hurley: 94 photographs, 120 pp.

Neff, W.D., 1978. Boundary-layer research at South Pole Station using acoustic remote sensing. A.J.U.S. 13 (4): 179-181.

Neff, W.D., and Hall, F.F., 1976a. Acoustic Sounder Measurements of the South Pole Boundary-layer. A.M.S. 17th Conference on Radar Meteorology: 297-302.

Neff, W.D., and Hall, F.F., 1976b. Acoustic echo sounding of the atmospheric boundary-layer at the South Pole. A.J.U.S. 11 (3): 143-144.

Neff, W.D., Ramm, H.E., and Hall, F.F., 1977. Acoustic sounder operations at South Pole Station. A.J.U.S. 12 (4): 167-168.

Neff, W.D., and Hall, F.F., 1978. Acoustic remote sensing of the planetary boundary-layer at the South Pole. A.M.S. Preprint Volume, Fourth Symposium on met. obs. and instrumentation, April 1978: 357-361.

Nordenskjoeld, O., 1911. Wissenschaftliche Ergebnisse der Schwedischen Suedpolar-Expedition 1901-1903. Lithographisches Institut des Generalstabs, Stockholm, Vol. I (1), 232 p.

Ohtake, T., 1976. Source of nuclei of atmospheric ice crystals at the South Pole. A.J.U.S. 11 (3): 148-149.

Ohata, T., Ishikawa, N., Kobayashi, S., and Kawaguchi, S., 1983. Micrometeorological data at Mizuho Station, Antarctica, in 1980. National Institute for Polar Research, Tokyo, JARE DATA REPORTS 79 (Met. 13): 371 pp.

Orheim, O., 1970. Volcanic activity on Deception Island, South Shetland Islands. Antarctic Geology and Geophysics, Symposium Oslo 1970, R.J. Adie (editor), Universitetet-forlaget Oslo 1972: 117-120.

Paltridge, G.W., and Platt, C.M.R., 1976. Radiative processes in meteorology and climatology. ELSEVIER SCI. PBL. CO.: 318 pp.

Parish, T.R., 1977. The flow of cold, stable air near the tip of the Antarctic Peninsula: an example of inertial flow. Research Report, Department of Meteorology, University of Wisconsin-Madison, 44 pp.

Parish, T., 1980. Surface winds in East-Antarctica. Ph.D. thesis and Research Report, Dept. of Meteorology, University of Wisconsin: 121 pp.

Parish, T.R., 1981. The katabatic winds of Cape Denison and Port Martin. Polar Record, 20 (129): 525-532.

Parish, T.R., 1982. Surface Airflow over East Antarctica. M.W.R. 110 (2): 84-90.

Parish, T.R., 1983. The influence of the Antarctic Peninsula on the wind field over the western Weddell Sea. J.G.R. 88 (C4): 2684-2692.

Parish, T.R., and Schwerdtfeger, W., 1977: A cold, low-level jet stream in the Bransfield Strait; an example of inertial flow. A.J.U.S. 12 (4): 171-172.

Peixoto, J.P., 1973. Atmospheric vapor flux computations for hydrological purposes. Reports on WMO/IHD Projects 20.

Pepper, I., 1954. The meteorology of the Falkland Islands and Dependencies 1944-50. Falkland Islands Dependencies Survey, London, 250 pp.

Peterson, H.C., 1948. Antarctic Weather Statistics compiled by the Ronne Antarctic Research Expedition. Office of Naval Research, Dpt. of the Navy, Technical Report 1, Washington, D.C.: 30 p.

Peterson, J.T., and Szware, V.S., 1977. Geophysical monitoring for climatic change at the South Pole. A.J.U.S. 12 (4): 159-160.

Petit, J.R., Jouzel, J., Pourchet, M., and Merlivat, L., 1982. A detailed study of snow accumulation and stable isotope content in Dome C, Antarctica. J.G.R. 87 (C6): 4301-4308.

Phillpot, H.R., 1967. Selected surface climatic data for Antarctic stations. Bureau of Meteorology, Melbourne, Australia. 113 pp.

Phillpot, H.R., 1968. A study of the synoptic climatology of the Antarctic. Internat. Antarctic Met. Research Centre, Melbourne, Australia. Technical Report 12: 139 pp.

Phillpot, H.R., and Zillman, J.W., 1970. The surface temperature inversion over the Antarctic continent. J.G.R., 75: 4161-4169.

Picciotto, E., Cameron, R., Crozaz, G., Deutsch, S., and Wilgain, S., 1968. Determination of the rate of snow accumulation at the Pole of Inaccessibility, East Antarctica: comparison of glaciological and isotopic methods. J. of Glac. 7 (50): 273-287.

Picciotto, E., Crozaz, E.G., and de Breuck, W., 1971. Accumulation on the South Pole - Queen Maud Land Traverse 1964-1968. A.G.U. Antarctic Res. Series 16: 257-315.

Polar Record, 1968. Volcanic Activity at Deception Island, South Shetland Islands. Vol. 14 (89): 229-230.

Polar Record, 1982. Stations operating in the Antarctic, winter 1982. Vol. 21 (132): 309-322.

Prandtl, L., 1942. Fuehrer durch die Stroemunglehre, pp. 373-375. Vieweg und Sohn, Braunshweig.

Priestley, R.E., 1915. Antarctic Adventure, Scott's Northern Party. New York E. P. Dutton & Co., 382 pp.

Priestley, R.E., Adie, R.J., and Robin, G. de Q., editors, 1964. Antarctic Research, a Review of British Scientific Achievement in Antarctica. London, Butterworths: 360 p.

Prohaska, F., 1951. Datos climatológicos y geomagnéticos, Islas Orcadas del Sur, período 1903-1950. Servicio Meteorológico Nacional, Bueno Aires, Publ. Serie B 1 (11): 97 pp.

Prohaska, F., 1954. Bemerkungen zum saekularen Gang der Temperatur im Suedpolargebiet. Arch. Meteorol., Geophys., Bioklim., Serie B 5: 327-330.

Prudhomme, A., and Le Quinio, R., 1954. Les observations météorologiques de Port-Martin en Terre Adélie. Fasc. II, Conditions atmospheriques en surface du 1. I. 1951 au 20. I. 1952. Relevés quotidiens. Expéd. Pol. Franc., Resultats Scientif., S 5:1 pp.

Prudhomme, A., and Valtat, B., 1957. Les observations météorologiques en Terre Adélie 1950-52, analyse critique. Expéditions Polaires Francaises, No. S V, 179 pp.

Pybus, J.A., and Brown Jr., J.A., 1967. Polar atmospheric water vapor. Proceedings of the Symposium on Polar Meteorology, Geneva 1966, WMO Technical Note No. 87: 3-14. Also see J.A.S. 1964, 21: 597-602.

Radok, U., 1977. Snow drift. J. of Glac. 19 (81): 123-129.

Radok, U., 1979. Polar meteorology and climatology 1975-78. Rev. Geophys. and Space Phys. 17: 1772-1781.

Radok, U., Schwerdtfeger, W., and Weller, G., 1968. Surface and subsurface meteorological conditions at Plateau Station. A.J.U.S. 3 (6): 257-258.

Radok, U., and Lile, R.C., 1977. A year of snow accumulation at Plateau Station. Paper 2 in Meteorological Studies at Plateau Station, A.G.U. Antarctic Research Series, Vol. 25, (J.A. Businger, editor): 17-26.

Reed, R.J., 1979. Cyclogenesis in Polar Air Streams. M.W.R. 107 (1): 38-52.

Reiter, E.R.., 1963. Jetstream Meteorology. University of Chicago Press: 515 pp.

Renard, R.J., and Salinas, M.G., 1977. The history, operation, and performance of an experimental automatic weather station in Antarctica. Technical Report NPS 63 Rd 77101 Naval Postgraduate School, Monterey, Cal.

Rey, J.J., 1911. "Météorologie". In: Exp. Antarc. Franc. 1903-05, Hydrographie, Physique du globe. Paris: 263-573.

Riordan, A.J., 1975. The climate of Vanda Station, Antarctica. In: Climate of the Arctic, 24th Alaska Science Conference August 1973 (G. Weller and Bowling, S.A., editors), Geophysical Institute, Univ. of Alaska, Fairbanks: 268-275.

Riordan, A.J., 1977. Variations of temperature and air motion in the 0 to 32 meter layer at Plateau Station. Paper 8 in Meteorological Studies at Plateau Station, A.G.U. Antarctic Research Series, Vol. 25 (A.J. Businger, editor): 113-127.

Robin, G. de Q., 1972. Polar ice sheets: a review. Polar Record 16 (100): 5-22.

Ronne, F., 1979. Antarctica my Destiny. Hastings House Publishers, New York, 278 pp.

Rouch, J., 1914. "Observations météorologiques". In: Deuxieme Exp. Antarc. Franc. 1908-1910, Sciences Physiques, Documents Scient., Paris.

Rubin, M.J., and Weyant, W.S., 1963. The mass and heat budget of the Antarctic atmosphere. M.W.R. 11 (10-12): 487-493.

Rusin, N.P., 1961: Meteorological and radiational regime of Antarctica. Translated from Russian by the Israel Program for Scientific Translations 1964 for DOC and NSF. 355 pp.

Sasaki, H., 1979. Preliminary study on the structure of the atmospheric surface layer in Mizuho Plateau, East Antarctica. Japan. Weather Assoc., Antarctic Record 67: 86-100.

Savage, M.L., and Stearns, C.R., 1981. Automatic weather stations in the Antarctic. In: Antarctic Weather and Climate, Preprint volume of the Univ. of Melbourne, Australia, May 1981: 5, 1-8.

Schatz, H., 1951. Ein Foehnstorm in Nordost-Groenland. Polarforschung, 3: 13-14.

Schneider-Carius, K., 1955. Wetterkunde-Wetterforschung, Geschichte ihrer Probleme und Erkenntnisse in Dokumenten aus drei Jahrtausenden. Verlag AlberFreiburg/Muenchen: 423 pp.

Schneider, O., 1969. La Barrera de Hielo de Filchner en el cincuentenario de su descubrimiento. Contribuciones del Instituto Antártico Argentino No. 63: 44 pp.

Schwerdtfeger, W., 1960. The seasonal variation of the strength of the southern circumpolar vortex. M.W.R. 88 (6): 203-208.

Schwerdtfeger, W., 1967. Annual and semi-annual changes of atmospheric mass over Antarctica. J.G.R. 72 (14): 3543-3547.

Schwerdtfeger, W., 1968. New data on the winter radiation balance at the South Pole. A.J.U.S. 3 (5): 193-194.

Schwerdtfeger, W., 1970a. The Climate of the Antarctic. Chapter 4, Volume XIV (S. Orvig, editor), World Survey of Climatology (H.E. Landsberg, editor-in-chief), Elsevier, Amsterdam, 253-355.

Schwerdtfeger, W., 1970b. Die Temperaturinversion ueber dem antarktischen Plateau und die Struktur ihres Windfeldes. Meteorologische Rundschau, 23(6): 164-171.

Schwerdtfeger, W., 1971. Remarkable wind shifts and speeds a few meters above
    the surface of the Antarctic Plateau. A.J.U.S. 6 (5): 218-219.
Schwerdtfeger, W., 1972. The vertical variation of the wind through the
    friction-layer over the Greenland ice cap. TELLUS 24 (1), 13-16.
Schwerdtfeger, W., 1974. Mountain barrier effect on the flow of stable air
    north of the Brooks Range. In: Climate of the Arctic, Proceedings of the
    24th Alaska Science Conference, 1973, (G. Weller and S.A. Bowling, editors):
    204-208.
Schwerdtfeger, W., 1975. The effect of the Antarctic Peninsula on the
    temperature regime of the Weddell Sea. M.W.R. 103 (1): 45-51.
Schwerdtfeger, W., 1976. Changes of Temperature and Ice Conditions in the Area
    of the Antarctic Peninsula. M.W.R. 104 (11): 1441-1443.
Schwerdtfeger, W., 1977. Mountain barrier winds in antarctic regions.
    Collection of Contributions presented at CPM Sessions, IAGA/IAMAP Assembly
    Seattle, Washington, September 1977, 84-100, M. Kuhn, editor, published by
    NCAR, 1979.
Schwerdtfeger, W., 1979. Meteorological aspects of the drift of ice from the
    Weddell Sea toward the mid-latitude westerlies. J.G.R. 84 (C 10): 6321-6328.
Schwerdtfeger, W., 1981. Elevation of Amundsen-Scott South Pole Station:
    2835 m. A.J.U.S. 16 (4): 10-11.
Schwerdtfeger, W., and Prohaska, F., 1955. Análisis de la marcha anual de la
    presión y sus relaciones con la circulación atmosférica en Sud América
    austral y la Antártida. Meteoros, 5: 223-237.
Schwerdtfeger, W., and Prohaska, F., 1956. Der Jahresgang des Luftdrucks auf
    der Erde und seine halbjaehrige Komponente. Meteorologische Rundschau,
    9: 33-43 and 186-187.
Schwerdtfeger, W., Canal, De La L.M., and Scholten, J., 1959. Meteorología
    Descriptiva del Sector Antártico Sudamericano. Publicaciones del Instituto
    Antártico Argentino, 7: 425 p.
Schwerdtfeger, W., and Martin, D.W., 1964. The zonal flow of the free atmo-
    sphere between 10°N and 80°S, in the South-American sector. J.A.M. 3 (6):
    726-733.
Schwerdtfeger, W., and Mahrt, L.J., 1968. The relation between terrain features,
    thermal wind, and surface wind over Antarctica. A.J.U.S. 3 (5): 190-191.
Schwerdtfeger, W., and Mahrt, L., 1970. The Relation between the Antarctic
    Temperature Inversion in the Surface-Layer and its Wind Regime. In:
    Proceedings of the International Symposium on Antarctic Glaciological
    Exploration, Hanover, N.H., September 1968, Publ. No. 86 of the Inter-
    national Association of Scientific Hydrology: 308-315.
Schwerdtfeger, W., and Komro, F., 1978. Early winter storms in the northwestern
    Weddell Sea. A.J.U.S. 13 (4): 175-177.
Schwerdtfeger, W., and Amaturo, L.R., 1979. Wind and Weather around the
    Antarctic Peninsula. Research Report, Dept. of Meteorology, University of
    Wisconsin, 86 pp.
Seilkopf, H., 1939. Maritime Meteorologie. Handbuch der Fliegerwetterkunde,
    II, Radetzki Verlag Berlin, 150 pp.
Shackleton, E.H., 1909. The heart of the Antarctic. London, William
    Heinemann: 2 vol.
Shaw, P.J.R., 1962. A note on the prevalence of the katabatic wind in
    Antarctica. Austral. Met. Mag. 36: 41-42.
Silverstein, S.C., 1967. The American Antarctic Mountaineering Expedition.
    Antarctic Journal of the United States, 2 (2): 48-50.
Simpson, G.C., 1919. British Antarctic Expedition 1910-13, Meteorology, Vol. I,
    Discussion. Thacker and Spink, Calcutta, 326 pp.
Sinclair, M.R., 1981. Record high temperatures in the Antarctic - a synoptic
    case study. M.W.R. 109 (10): 2234-2242.
Siple, P., 1959. 90°SOUTH, The story of the American South Pole conquest.
    Putnam's, New York, 384 pp.

Siple, P.A., and Passel, C.F., 1945. Measurements of dry atmospheric cooling in subfreezing temperatures. Proc. Am. Phil. Soc. 89: 177-199.

Sissala, J.E., Sabatini, R.R., and Ackermann, H.J., 1972. NIMBUS Satellite Data for Polar Ice Survey. Polar Record 16 (102): 367-373.

Skeib, G., 1963. Meteorological investigations at temporary station MIR. S.A.E.I.B. 37: 207-210.

Smiley, V.N., 1980. Lidar measurements in the Antarctic. A.J.U.S. XV (5): 188-190.

Smiley, V.N., Warburton, J.A., and Morley, B.M., 1975. South Pole ice crystal precipitation studies using lidar sounding and replication. A.J.U.S. 10 (5): 230-231.

Smiley, V.N., Warburton, J.A., Morley, B.M., and Whitcomb, B.M., 1976. Lidar studies for polar regions. A.J.U.S. 11 (3): 145-148.

Smiley, V.N., Morley, B.M., and Warburton, J.A., 1977. Lidar and replication studies of ice crystal precipitation at the South Pole. A.J.U.S. XII (4): 166-168.

Smiley, V.N., and Morely, B.M., 1981. Lidar depolarization studies of the atmosphere at the South pole. Appl. Optics 20: 2189-2195.

Solot, S.B., 1967. GHOST Atlas of the Southern Hemisphere. NCAR Publ.: 10 pp.

Solot, S.B., 1968. GHOST Balloon Data. NCAR Techn. Note 34: 11 volumes, 1729 pp.

Solot, S.B., and Angell, J.K., 1969. Mean meridional air flow implied by 200 mb GHOST balloon flights in temperate latitudes of the Southern Hemisphere. J.A.S. 26 (6): 1299-1305.

South African Weather Bureau, 1962-66. International Geophysical Year. World Weather Maps, Part III: Southern Hemisphere south of 20°S. Daily sea level and 500 mb charts, July 1957 - December 1958.

Sponholz, M., and Schwerdtfeger, W., 1970. Theory and observations of the wind in the friction-layer over the Antarctic Plateau. A.J.U.S. 5 (5): 175-176.

Stearns, C.R., and Savage, M.L., 1981. Automatic weather stations, 1980-1981. A.J.U.S. 16 (5): 190-192.

Stearns, C.R., 1982. Antarctic automatic weather stations, 1982-1983. A.J.U.S. 17 (5): 217-219.

Stephenson, A., 1938. Notes on the Scientific Work of the British Graham Land Expedition 1934-37, Meteorology. The Geographical Journal London, 91: 518-523.

Streten, N.A., 1962. Note on weather conditions in Antarctica. Austral. Met. Mag. 37: 1-20.

Streten, N.A., 1963. Some observations of antarctic katabatic winds. Austral. Met. Mag. 42: 1-23.

Streten, N.A., 1975. Satellite derived inferences to some characteristics of the South Pacific atmospheric circulation associated with the Niño event of 1972-73. M.W.R. 103: 989-995.

Streten, N.A., 1981. A Prospect of Antarctic Meteorology - The Australian Experience. In: Antarctica, Weather and Climate, Preprint volume compiled by N.W. Young, University of Melbourne, May 1981, 7 pp.

Streten, N.A., and Kellas, W.R., 1973. Aspects of cloud signatures of depressions in maturity and decay. J.A.M. 12 (1): 23-27. Also see Austral. Met. Mag. 23 (1): 7-19, 1975.

Streten, N.A., and Troup, A.J., 1973. A synoptic climatology of satellite observed cloud vortices over the Southern Hemisphere. Quart. Journ. Roy. Met. Soc. 99: 56-72.

Streten, N.A., and Pike, D.J., 1980. Indices of the mean monthly surface circulation over the southern hemisphere during FGGE. Austral. Met. Mag. 28 (4): 201-216.

Swanson, G.S., and Trenbreth, K.E., 1981. Trends and interannual variability in the Southern Hemisphere tropospheric circulation. M.W.R. 109: 1879-1897.

Swithinbank, C., 1973. Higher resolution satellite pictures. Polar Record 16 (104): 739-741.

Swithinbank, C., McClain, P., and Little, P., 1977. Drift tracks of
Antarctic icebergs. Polar Record 18 (116): 495-501.

Takeuchi, M., 1980. Vertical profile and horizontal increase of drift-snow
transport. J. of Glac. 26 (94): 481-492.
Taljaard, J.J., 1972. Synoptic Meteorology of the Southern Hemisphere. Part of
Am. Met. Soc. Meteorological Monographs 35, C.W. Newton, editor: 139-213.
Taljaard, J.J., Schmitt, W., and van Loon, H., 1961. Frontal analysis with
application to the Southern Hemisphere. NOTOS, 10: 25-58.
Taljaard, J.J., and van Loon, H., 1962 and 1963. Cyclogenesis, cyclones and
anticyclones in the Southern Hemisphere during the winter and spring of
1957; and the same title, but during summer 1957-58. NOTOS 11: 3-20, and
NOTOS 12: 37-50.
Taljaard, J.J., and van Loon, H., 1964. Southern Hemisphere weather maps for
the I.G.Y. B.A.M.S. 45: 88-95.
Taljaard, J.J., van Loon, H., Crutcher,, H.L., and Jenne, R.L., 1969. Climate
of the upper air, Part 1 - Southern Hemisphere, Vol. 1, Temperature, dew
points, and heights at selected pressure levels. NAVAIR 50-1C-55 Washington,
D.C., 135 p.
Tauber, G.M., 1960. Characteristics of Antarctic katabatic winds. In:
Antarctic Meteorology, Proceedings of the Symposium held at Melbourne 1959,
Pergamon Press, Oxford: 52-64.
Taylor, G.I., 1916. Skin friction of the wind on the earth's surface.
Proc. Roy. Soc. London, A 92: 196-199.
Tchernia, P., 1974. Etude de la dérive antarctique est-ouest au moyen
d'icebergs suivis par le satellite Eole. C.R. Acad. Sci. Paris 278: 667-670.
Tchernia, P., and Jeannin, P.F., 1980. Observations on the Antarctic east wind
drift using tabular icebergs tracked by satellite Nimbus F, 1975-77.
Deep-Sea Research Part A 27 (6A) June 1980.
Thomas, R.H., 1976. The distribution of 10 m ice temperatures on the Ross Ice
Shelf. Journ. of Glac. 16 (74): 111-117.
Thomas, R.H., Sanderson, T.J.O., and Rose, K.E., 1979. Effect of climatic
warming on the west-antarctic ice sheet. NATURE, Febr. 1, 1979, 277 (5695):
355-358.
Thompson, D.C., and MacDonald, W.J.P., 1961. Meteorology -- Scott Base.
New Zealand Meteorological Service Met. Office Note 48: 37-58.
Thompson, D.C., and MacDonald, W.J.P., 1962. Radiation measurements at Scott-
Base. New Zealand Journal of Geology and Geophysics, 5: 874-909.
Thompson, D.C., Craig, R., and Bromley, A., 1971. Climate and surface heat
balance in an Antarctic Dry Valley. Technical Note 193, New Zealand Met.
Service: 20 p.
Trenberth, K.E., and van Loon, H., 1981. Comment on "Impact of FGGE Buoy Data
on Southern Hemisphere Analyses". B.A.M.S. 62 (10): 1486-1488.
Treshnikov, A.F., 1967. The ice in the Southern Ocean. In: Proceedings of
the Symposium on Pacific-Antarctic Sciences. Eleventh Pacific Science
Congress, Tokyo 1966: 113-123.

Ueda, H.T., and Garfield, D.E., 1968. Deep-core drilling program at Byrd
station (1967-68). A.J.U.S. 3 (4): 111-112.
Ugolini, F.C., and Bull, C., 1965. Soil development and glacial events in
Antarctica. Quaternaria VII: 251-269.
U.S. Naval Polar Oceanography Center, 1971-present. Weekly ice maps around
Antarctica.
U.S. Navy Fleet Weather Facility, 1971-79. Earlier name of U.S. Naval Polar
Oceanography Center, Suitland.
U.S. Navy, 1965. Marine Climatic Atlas of the World, Vol. 7, Antarctica.
NAVWEPS 50-1C-50, U.S. Gvt. Printing Office, Washington, D.C.: 361 p.

U.S. Weather Bureau and WMO, 1956-82. Monthly Climatic Data for the World, sponsored by the World Meteorological Organization in cooperation with the U.S. Weather Bureau (ESSA, later NOAA). Volumes 9 through 35, 1956-82.

U.S. Weather Bureau and successors (ESSA and later NOAA), 1962-1977. Climatological Data for Antarctic Stations (U.S. stations only) for the years 1957 to 1975, issues 1 to 14. U.S. Government Printing Office, Washington, D.C.

van Loon, H., 1962. On the movement of Lows in the Ross and Weddell Sea Sectors in Summer. NOTOS, 11: 47-50.

van Loon, H., 1964. Mid-season average zonal winds at sea level and at 500 mb south of 25°S, and a brief comparison with the Northern Hemisphere. J.A.M. 3: 554-563.

van Loon, H., 1966. On the annual temperature range over the Southern Oceans. The Geographical Review, 56 (4): 495-515.

van Loon, H., 1967. The half-yearly oscillations in middle and high Southern latitudes and the coreless winter. J.A.S. 24(5): 472-486.

van Loon, H., 1972. Temperature in the Southern Hemisphere, and Wind in the Southern Hemisphere. Chapters 3 and 5 in Meteorology of the Southern Hemisphere, C.W. Newton (editor). Met. Mon. A.M.S. 13(35): 25-58 and 87-100.

van Loon, H., Taljaard, J.J., Jenne, R.L., and Crutcher, H.L., 1971. Climate of the Upper Air: Southern Hemisphere, Vol. II. NAVAIR 50-1C-56 and NCAR TN/STR-57, Boulder Colorado: 39 pp.

van Loon, H., and Jenne, R.L., 1974. Standard deviations of monthly mean 500 and 100 mb heights in the southern hemisphere. J.G.R. 79 (36): 5661-5664.

van Loon, H., and Williams, J., 1977. The connection between trends of mean temperature and circulation at the surface: Part IV. Comparison of the surface changes in the northern hemisphere with the upper air and with the Antarctic in winter. M.W.R. 105 (5): 636-647.

van Loon, H., and Rogers, J.C., 1981. Remarks on the circulation over the southern hemisphere in FGGE and on its relation to the phases of the Southern Oscillation. M.W.R. 109 (11): 2255-2259.

Venter, R.J., 1957. Sources of meteorological data for the Antarctic. Chapter 2 in Meteorology of the Antarctic, M.P. van Rooy, editor, Weather Bureau of South Africa, Pretoria: 17-38.

Vickers, W.W., 1966. A study of ice accumulation and tropospheric circulation in western Antarctica. In: Studies in Antarctic Meteorology, Antarctic Research Series Vol. 9 (M.J. Rubin, editor): 135-176.

Viebrock, H.J., and Flowers, E.C., 1968. Comments on the recent decrease in solar radiation at the South Pole. TELLUS, 20 (3): 400-411.

Vinje, T.E., 1965. Climatological tables for Norway station, Instruments and surface observations. Norske Antarktisekspedisjonen 1956-60, Scientific Results No. 8, Norsk Polarinstitutt, Årbok 1963, pp. 181-183.

Vinje, T.E., 1976. Drift av Trolltunga i Weddellhavet. Norsk Polarinstitutt Årbok 1975: 213.

Vinje, T.E., 1979. On the drift-ice conditions in the Atlantic sector of the Antarctic. Internat. Conference on Port and Ocean Engineering under Arctic Conditions, 5th, Trondheim, Norway, 13-18 Aug. 1979, Proceedings Vol. 3, 75-82, 1979.

Vinje, T.E., 1980. Some satellite-tracked iceberg drifts in the Antarctic. Annals of Glaciology 1: 83-87.

Vinje, T.E., 1981. Meteorological observations from Bouvetøya. Norsk Polarimstitutt Skrifter Nr. 175: 85-103.

Vowinckel, E., and Orvig, S., 1970. The Climate of the North Polar Basin. In: Climate of the Polar Regions (S. Orvig, editor), Vol. 14 of World Survey of Climatology (H.E. Landsberg, editor i C.): 129-252. Also see P. Putnins, same volume: 3-128.

Wada, M., Yamanouchi, T., Mae, S., Kawaguchi, S., and Kusunoki, K., 1981.
    Micrometeorological data at Mizuho Station, Antarctica, in 1979.
    National Institute of Polar Research, Tokyo. JARE DATA REPORTS 62 (Met. 9):
    321 pp.
Warren, S.G., 1982. Optical properties of snow. Reviews of Geophysics and
    Space Physics, 20 (1): 67-89. Note: The same issue of R.G.S.P. contains
    five other papers on the physics of snow.
Weddell, J., 1827. A voyage towards the South Pole. (A reprint of the second
    edition, 1827, with an introduction by Vivian Fuchs, published 1970). David
    and Charles, Publishers, England. 324 p.
Wegener, A., 1911. Thermodynamik der Atmosphäre. Verlag J.A. Barth, Leipzig
    (third edition, 1928): 331 pp.
Weickmann, H., 1981. Mechanism of shallow winter-type stratiform cloud systems.
    Publication of NOAA, Environmental Research Labs. at Boulder, October 1981:
    61 pp.
Weller, G.E., 1968. The heat budget and heat transfer processes in Antarctic
    Plateau ice and sea ice. ANARE Scientific Reports, Series A (IV) Glaciology,
    Publ. No. 102, Melbourne, Australia, 155 pp.
Weller, G., 1970. Meridional surface wind speed profile in MacRobertson Land,
    Antarctica. Pure and Applied Geophysics, 77: 193-200.
Weller, G., 1980. Spatial and temporal variations in the south polar energy
    balance. M.W.R., 108 (12): 2006-2014.
Whillans, I.M., 1975. Effect of inversion winds on topographic detail and
    mass balance on inland ice sheets. J. of Glac. 14 (70): 85-90.
Wilkinson, F.W.A., 1967. Adelaide Piedmont Wind. Preliminary Report 1966/67,
    16 pp. British Antarctic Survey, Courtesy Mr. D.W.S. Limbert.
Wilson, E., 1972. Diary of the TERRA NOVA Expedition to the Antarctic 1910-12.
    Edited by H.G.R. King, Humanities Press, New York.

Yamanouchi, T., Wada, M., Mae, S., and Kawaguchi, S., 1981. Radiation Data at
    Mizuho Station, Antarctica, in 1979. JARE Data Reports, National Institute
    of Polar Research, Tokyo, 61: 350 pp.
Yamanouchi, T., Wada, M., Mae, S., Kawaguchi, S., and Tsukamura, K., 1981.
    Measurements of radiation components at Mizuho Station, East Antarctica,
    in 1979. Memoirs of the National Institute of Polar Research, Tokyo.
    Special Issue No. 19: 27-39.

Zavialova, I.N., 1980. Atmospheric humidity regime in Antarctica. In: Climatic
    research in the Antarctic, I.M. Dolgina, editor. Hydrometeorological
    Institute, 190 pp: 90-99. In Russian.
Zillman, J.W., 1967. The surface radiation balance in high southern latitudes.
    Proceedings of the Symposium on Polar Meteorology, Geneva, 1966. WMO-
    Technical Note No. 87: 142-174.

258

SUBJECT INDEX

Ablation, 197
Ablation areas, 205
Accumulation, 197-202
 definition, 197
 methods of measurement, 198
 new data for SOUTH POLE, 199
 on the high plateau, 200
 on Ross Ice Shelf, 201
 units, 199
 variations at BYRD, 199
Acoustic sounding, 10, 35
Advection, thermal,, 182
Advective heating, 14, 30, 125-128
Air masses, 159-161
Albedo, 14, 20, 22
Allwave net radiation, 13, 25
Ångström ratio, 15, 21, 23, 29
Antarctic Sound, 106
Anticyclones, 156-158
 blocking, 156
 highest pressure values
  recorded, 144
 migrating, 157
 over the high plateau, 164-167
Atmospheric radiation, 15, 23
Automatic weather stations, 8, 40,
 61, 82, 95-98

Barrier winds, 42, 78-99
 on Larsen Ice Shelf, 86-88
 related to Foehn winds, 90-93
 on Ross Ice Shelf, 94-98
 width of stream, 98
Beardmore Glacier, 26, 96, 154, 201
Blizzards, 94, 114, 116
Blowing snow, 61, 114-117
Bora winds, 41, 100
Bransfield Strait, 80, 106
Bullets (ice), 190
Buoys of FGGE, 9-10, 129

Calms, 41, 113
CAPE ADARE, 113, 114
 ARMITAGE, 39
 CROZIER, 94
 KEELER, 80, 92
Carbon dioxide, increase of
 concentration, 216
Central Polar Scientific Archives, 90
Circulation, zonal and meridional,
 120-135
Circumpolar low pressure trough,
 139, 142
Circumpolar vortex, 120-135
Clausius-Clapeyron equation, 170

Clean Air Facility, 6
Clear Sky precipitation on the
 plateau, 188-194
Climate divide, 79, 81, 87
Climate variations, 212-232
Cloudiness, 23-25, 178-186
 effect on longwave radiation, 23-25
 frequency of clear and cloudy sky,
  179
 frequency of cloud types, 180
 mean of low cloudiness, 178, 183
 mean of total cloudiness, 178
Columnar ice crystals, 190
Comparison, north vs. south polar
 regions, 223-227
 mean circumpolar pressure field, 224
 westerly winds in upper troposphere,
  224
 moisture content, 224
 occurrence of sudden stratospheric
  warmings, 224
 temperature of troposphere at 80°N
  and 80°S, 225
 of VERKHOYANSK and VOSTOK, 226
 annual, 80 years, MURMANSK and
  ORCADAS, 227
Continental antarctic air, 160
Convergent fields of motion, 76
Cooling rates in the lower troposphere,
 37, 125-127
Cooling of the South Atlantic Ocean,
 212
Coreless winter, 26
CREST Station, 80, 92
Crossing of low level streams of air,
 109
Cyclones, 136-155
 effect on katabatic winds, 58,, 65,
  68
 lowest pressure values in center,
  141-144
 migrating, 139
 of polar regions, 141
 over the high plateau, 165
 trajectories in Antarctic Peninsula
  region, 105

Damming up of cold stable airmasses, 84
DECEPTION ISLAND, volcanic eruptions, 6
Deserts, 3, 199
Deposition, 168, 194
Dew-point temperature, 168
Directional constancy, 42, 46, 86, 90
Diurnal variation of temperature at
 high plateau, 30, 32